KV-638-357

Knowledge Quest

Body and Health

The human body evolved over millions of years, but only in recent times have we come close to a full understanding of how this amazing machine works.

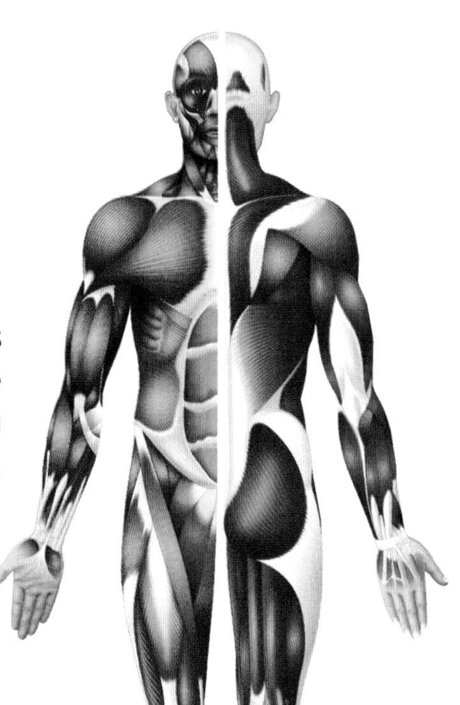

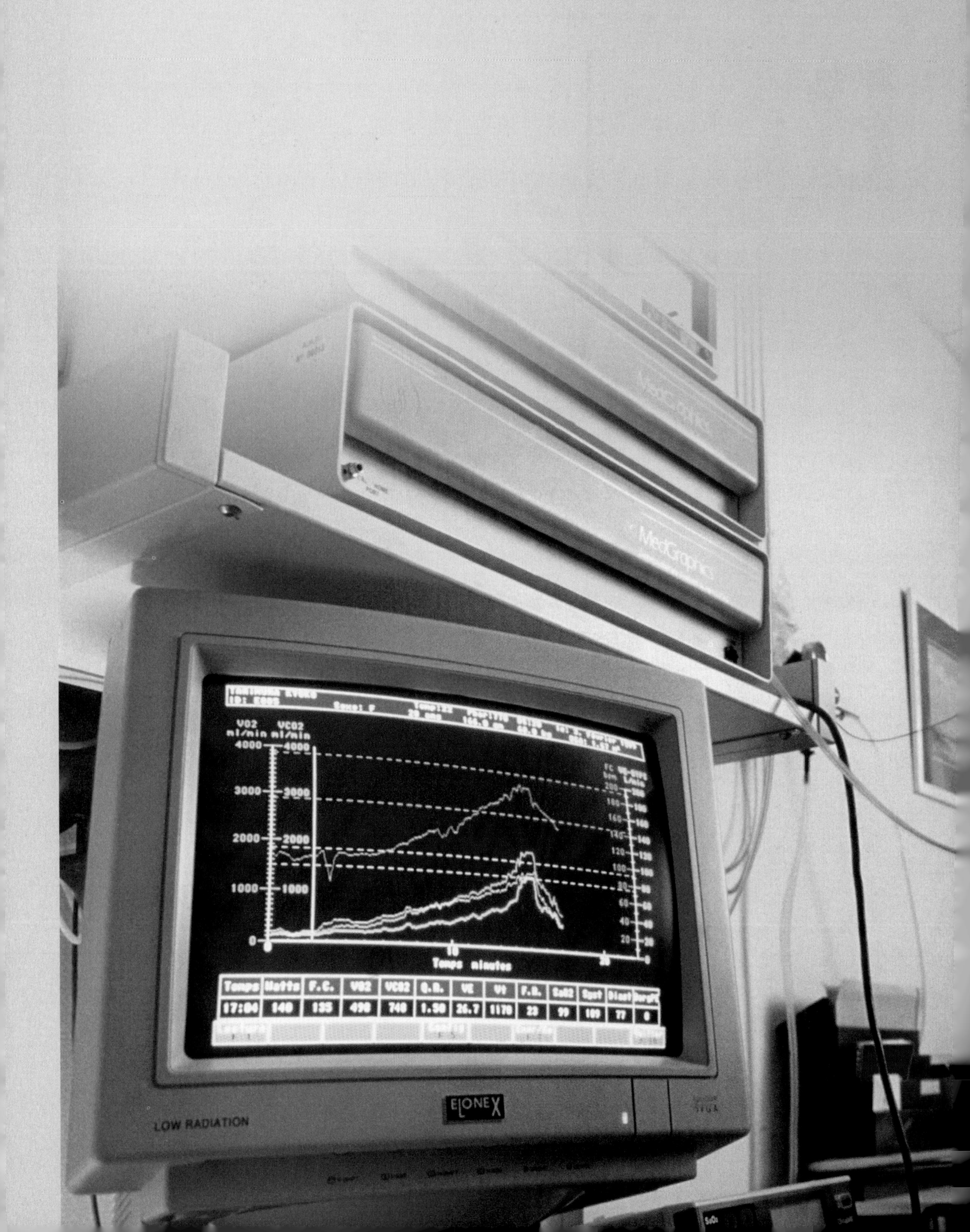

MedGraphics
VO2 VCO2
ml/min ml/min
Temps minutes
ELONEX
LOW RADIATION

Body and Health

Published by
The Reader's Digest Association Limited
London • New York • Sydney • Montreal

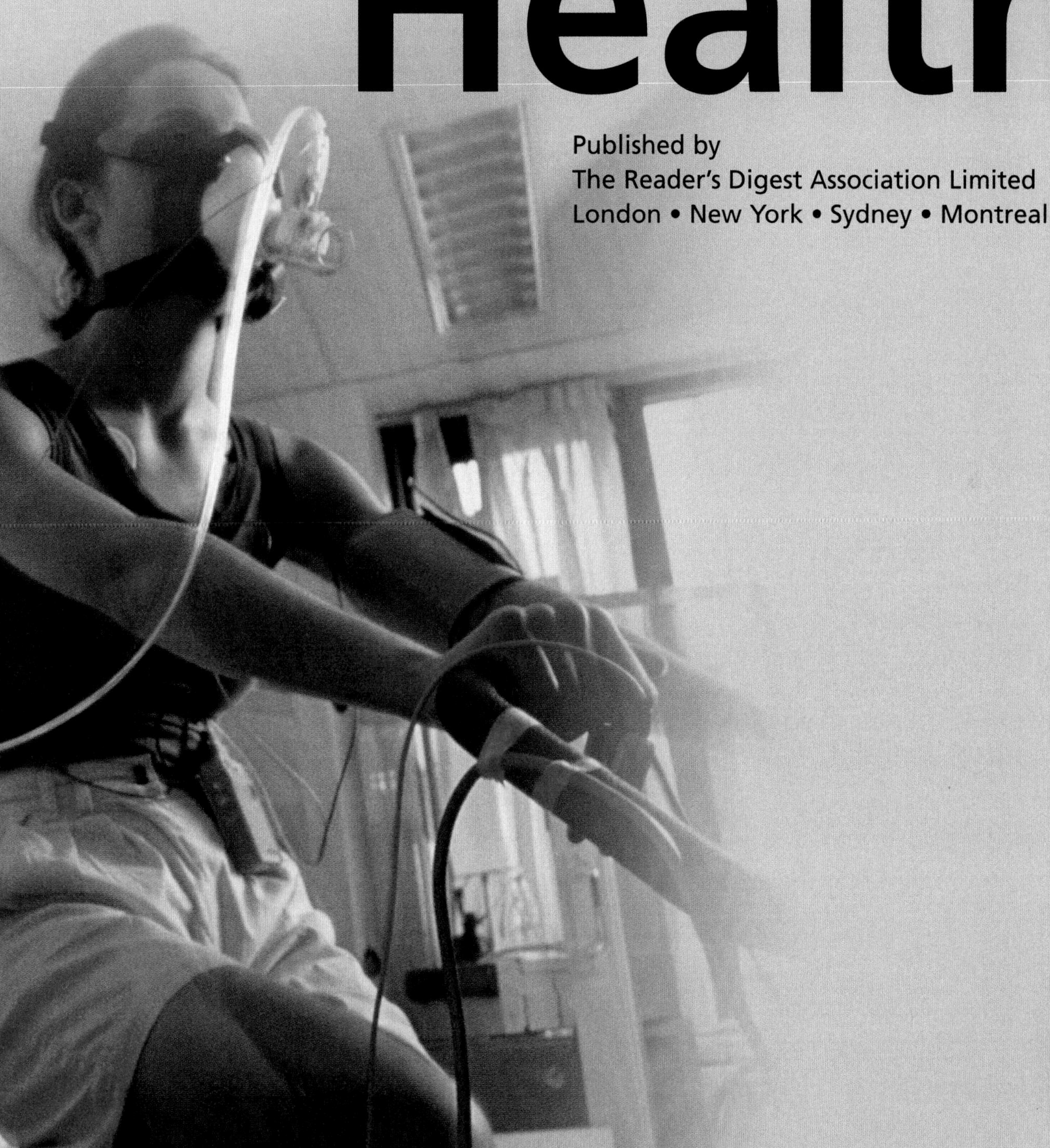

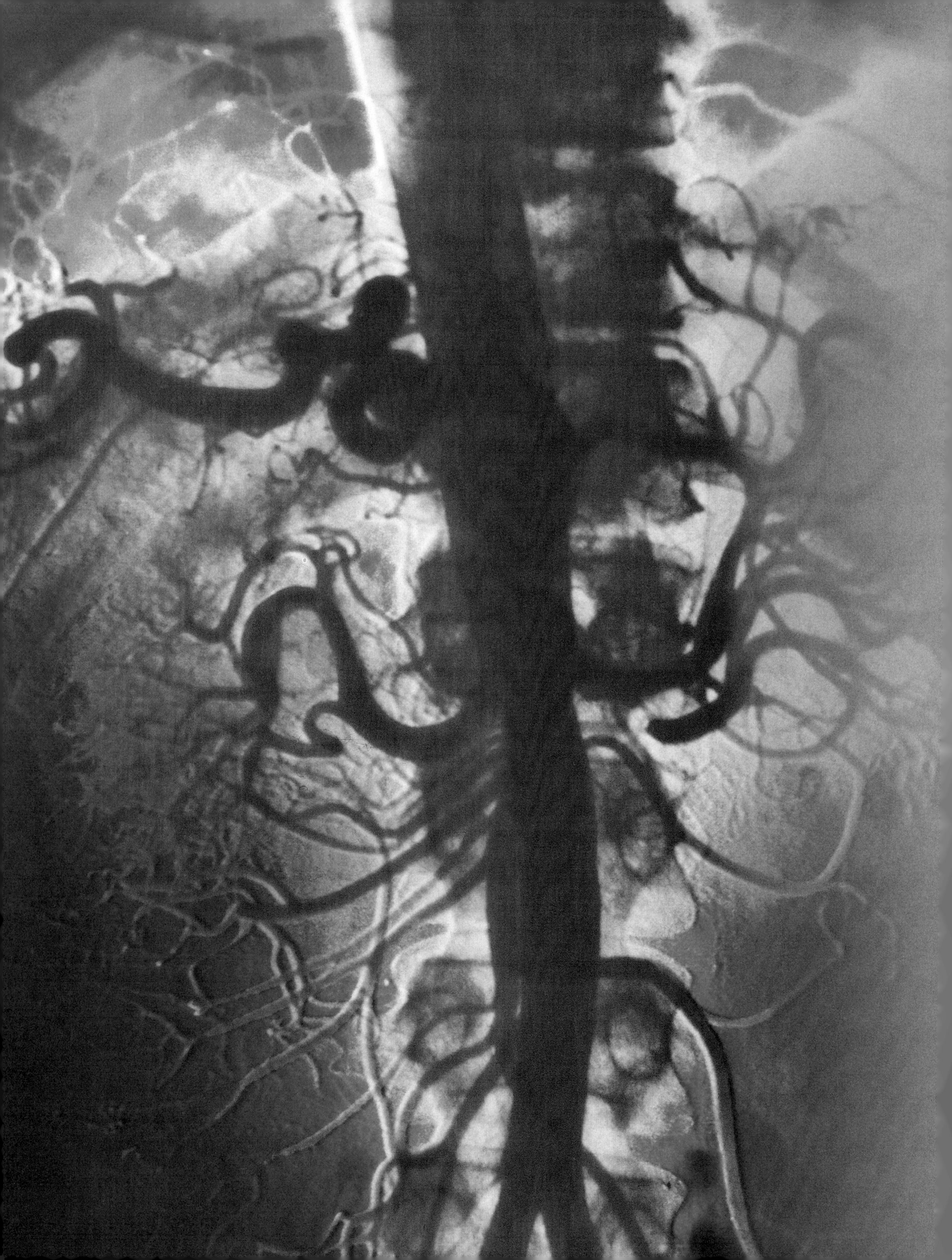

Contents

How to use Knowledge Quest

Knowledge Quest: Body and Health is a uniquely interactive reference book that brings you the essential facts and a wealth of information on the human body and medical procedures.

Knowledge Quest will make adding to your store of knowledge both interesting and fun. It builds into a highly illustrated reference series, with each volume delivering authoritative facts and many other significant things to know about a particular branch of knowledge.

Body and Health is packed with accessible facts and figures about the workings of the human body and branches of medicine, all contained within the core reference section (pages 39 to 145). If you want to use the book as a straightforward source of information the contents page lists the major topics covered, while the detailed index (starting on page 153) allows you to look up specific subjects. If you simply like to browse, you will find that each fascinating piece of information leads you to discover another, and another, and another . . .

What's so special about Knowledge Quest?

The unique feature of **Knowledge Quest** is the set of quiz questions that can be used as an entertaining way to get into the reference information. You can use these to test out what you already know about the human body, and to lead you eagerly into finding out more.

How do the questions link to the reference section?

There are 100 quizzes of 10 questions each, which are graded and colour-coded for levels of difficulty (see right). Each question is accompanied by a page number, which is the page in the reference section where you will find both the answer and more information on that subject generally.

The answers to all questions relating to the topic of the spread – which will come from several quizzes – are listed by question number in the far-left column. A number (or sometimes a star) following each answer refers you to the box containing the most relevant additional information elsewhere on the spread. More than one box may be indicated, and sometimes none are especially relevant, in which case an additional detail is given with the answer.

The questions in each quiz lead you to several different pages and topics in the reference section, which is great for finding out more, but is not so convenient if you want to use the quizzes as straightforward quiz rounds. So, for all keen quizmasters, we also list the answers to all the questions in each quiz in the 'Quick answers' section which starts on page 146.

Each question in the quizzes at the front of the book is linked by a page number (given immediately below the question number) to a page in the reference section, where you will find the answer, plus additional information on the subject. In this example, question 8 is linked to page 52.

The answers to all the questions relevant to the overall topic of a reference spread are listed by question number in the left-hand column. Each answer is followed by a box reference showing where more information can be found. Sometimes, instead of a box reference, additional information is given with the answer.

Each box on the rest of the reference spread contains information about an aspect of the overall topic. In this example, the topic of pages 52-53 is the heart and circulation and box ❺ provides information on how the body's circulatory system works. Sometimes, as here, more than one box will be relevant to an answer.

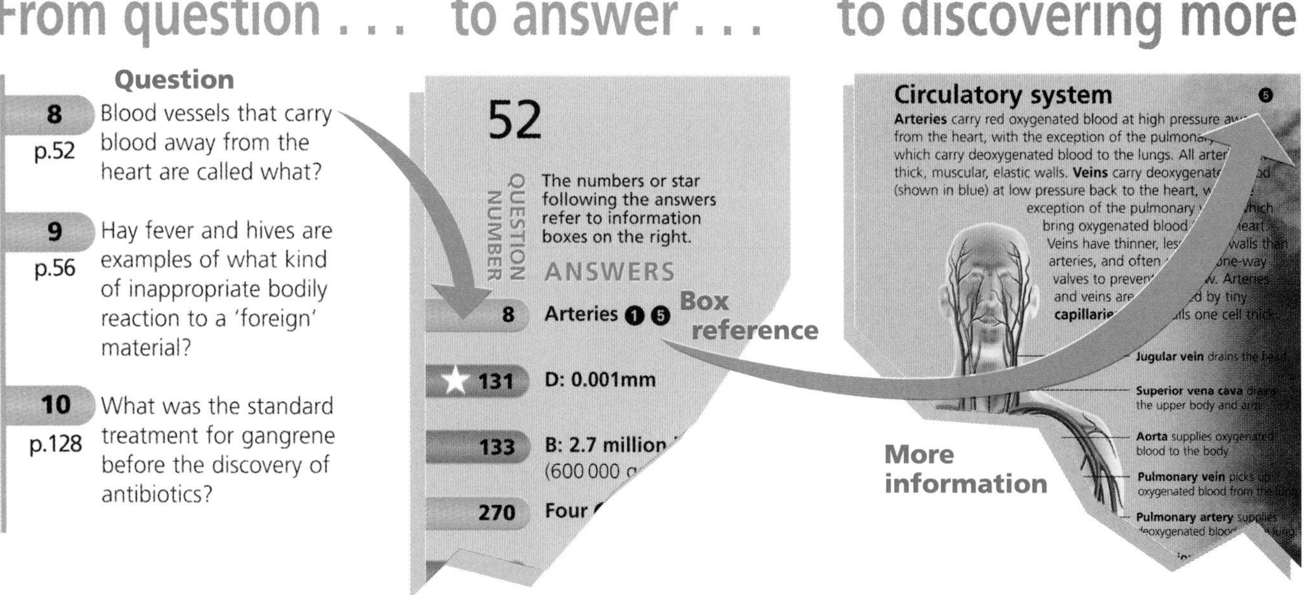

Colour-coding for quizzes of different levels of difficulty

A thousand questions are provided in a hundred themed quizzes. There are three levels of difficulty – **Warm Up**, **In Your Stride** and **Challenge**:

Warm Up quizzes feature easy questions to get you into the swing. Children may enjoy these quizzes as much as adults.

In Your Stride questions are pitched at a more difficult level than 'Warm Up' quizzes, requiring a little more knowledge of the subject.

Challenge quizzes are the most difficult, often requiring in-depth or specialist knowledge. But have a go anyway – you never know what information may be tucked away in the recesses of your brain!

Two other categories, **All Comers** and **Multiple Choice**, include mixed-level questions, ranging from 'warm-up' level to 'challenge', and generally becoming harder as a round progresses.

All Comers quizzes cover a range of levels of questions, from easy to hard. Everyone should be able to join in and see how they get on.

Multiple Choice quizzes offer four possible answers to each question, only one of which is correct. They feature mixed-ability questions, generally arranged to become harder as a quiz progresses.

Special features

Star answers

One answer on each two-page spread is marked with a **star** in the answer column. This indicates a subject of special or unusual interest within the spread topic. (Occasionally, more than one answer in the answers column refers the reader to the star answer box.)

The voice box

The voice box

The **larynx**, or voice box, is a hollow tube near the top of the trachea. On either side of the larynx, the **vocal cords** (two

Tie-breaker questions

Some spreads, but not all, feature tie-breaker questions:

Tie-breaker

These questions can be used by quizmasters in a quiz in the event of a tie, but they also contain information that expands on the answer for the interested reader.

Keeping score

The most straightforward **scoring method** is simply to award one point – or if you prefer, two points – for every correct answer.

If you are using the quizzes as rounds in a competition, you may find it easiest to look up the answers in the **Quick Answers** section at the back of the book.

For readers who would like to use this book as an information source for setting quizzes, a blank **Question sheet** and **Answer sheet** are provided at the back of the book. Quizmasters can photocopy the question sheet for their own use, and the answer sheet for distribution among contestants.

Give me an A

All the answers begin with the letter A.

1 p.120 What was the old name for a pharmacist?

2 p.66 What eating disorder mainly afflicts teenage girls who believe that they are overweight?

3 p.142 What is another common name for a post mortem examination?

4 p.128 For what purpose might a medieval surgeon have used brandy, mandrake or opium?

5 p.122 In what Chinese healing technique are needles inserted into certain points on the body?

6 p.124 What popular method of massage uses highly scented plant oils?

7 p.82 What hormone is produced by the adrenal glands when you are excited or scared?

8 p.52 Blood vessels that carry blood away from the heart are called what?

9 p.56 Hay fever and hives are examples of what kind of inappropriate bodily reaction to a 'foreign' material?

10 p.128 What was the standard treatment for gangrene before the discovery of antibiotics?

Doctorspeak

Test yourself on some medical terms.

11 p.142 What is the literal translation of *post mortem*?

12 p.130 What type of anaesthetic numbs only the area being operated on?

13 p.136 What device is used for injecting drugs into or under the skin (two words)?

14 p.128 What is the name (from French) for a strap or clamp used temporarily to stop bleeding from a limb?

15 p.74 The brain condition that may cause convulsions (fits) and unconsciousness is called what?

16 p.70 Many liver problems, such as hepatitis, cause yellowing of the skin and the whites of the eyes. What is this symptom called?

17 p.124 Osteopathy is one of two well-known forms of alternative manipulation therapy; What is the other one called?

18 p.130 Minimally invasive surgery is commonly called what?

19 p.142 What medical term describes the stiffening of muscles that sets in a few hours after death?

20 p.86 What term is used for a baby developing in the womb, from about the eighth week of pregnancy onwards?

True or false?

Decide whether these statements are correct.

21 p.76 Red-green colour blindness is much more common in men and boys than in women and girls.

22 p.96 Going without sleep for several days can cause hallucinations.

23 p.60 When a baby is born, the adult teeth are already forming in his or her jaws.

24 p.60 Humans are unique in having a voice box, thus allowing them to speak.

25 p.94 Your pulse rate gives a good clue to your fitness level.

26 p.74 If you touch a hot surface, you pull your hand away before the brain registers the discomfort.

27 p.116 In heart failure the heart stops beating.

28 p.98 Psychiatry is another word for psychology.

29 p.130 Brain surgery is usually performed while the patient is conscious.

30 p.76 The pattern of colours and lines on the iris of the eye is unique to each person.

Animal connections

Ten questions linked by an animal theme.

31 p.82 Which hormone-based drug, once extracted from the pancreas of animals, is today made by genetically modified bacteria?

32 p.116 *Cancer* is the Latin word for what kind of creature?

33 p.126 The term vaccination derives from which farm animal?

34 p.54 Which winged mammals are named for their habit of feeding on blood, occasionally transmitting rabies?

35 p.122 In which Eastern form of exercise would you come across the fish, the lion, the locust and the scorpion?

36 p.54 Which system of blood groups is named after the type of monkey used in experiments that uncovered it?

37 p.60 Incisors, premolars and molars are three types of adult teeth; name the fourth.

38 p.78 The hearing organ in the inner ear is shaped like a snail shell. What is it called?

39 p.124 What is the modern name for 'animal magnetism'?

40 p.112 What is the general medical term for insects and other creatures that spread disease by biting people?

For answers and more facts go to the page given below each question number.
For quick answers to complete quizzes 0 to 6 go to page 146.

Number crunch

A quiz round on numbers, weights and measures.

41 p.44 How many chromosomes do all human cells, except eggs and sperm, contain?

42 p.54 How many blood types are there in the combined ABO and Rhesus systems?

43 p.68 To the nearest metre (3 ft), how far is it from an adult's mouth to anus, following the path taken by food?

44 p.74 How many pairs of cranial nerves branch out from the brain?

45 p.74 How many pairs of spinal nerves branch out from the spinal cord?

46 p.84 Sex movies used to be rated X. What is the 'highest' current classification for adult-only films in Britain?

47 p.88 What percentage of babies are born prematurely: 5-10 per cent or 15-20 per cent?

48 p.90 To the nearest 0.1kg (1/4 lb), what is the average weight of a newborn baby?

49 p.90 To the nearest 5 cm (2 in), what is the average length of a newborn baby?

50 p.94 To calculate a person's maximum heart rate, you deduct a figure from 220: is this figure the person's age, or their weight in kilograms?

Name the therapy

All the answers are types of medical treatment.

51 p.108 What is the medical term for an exploratory abdominal operation?

52 p.110 What is the medical term for the surgical removal of a small sample of body tissue for testing or examination?

53 p.134 What is the name of the machine used to shock a heart-attack victim's heart back to a normal beat?

54 p.120 For what operation was a lancet widely used until the 19th century?

55 p.122 In what variation of acupuncture are hot needles or small burning cones of moxa applied to the skin?

56 p.124 What is the system of medicine, founded by Samuel Hahnemann, based on the idea that 'like cures like'?

57 p.134 TeNS – short for transcutaneous electrical nerve stimulation – is used to treat what?

58 p.136 What does the word 'hypodermic' mean?

59 p.138 In relation to medication, what does 'topical' mean to a doctor or pharmacist?

60 p.138 What common condition is treated with ACE inhibitors, beta blockers or calcium-channel blockers?

Mixed bag

Select the correct option from the four possible answers.

61 p.72 What are the membranes covering the brain and spinal cord called?

A Cerebrum | B Cortex | C Meninges | D Skull

62 p.108 Which imaging technique shows up brain activity?

A CT or CAT scan | B MRI | C PET scan | D Ultrasound

63 p.108 Which of these medical imaging techniques uses X-rays?

A CT or CAT scan | B MRI | C Radionuclide scanning | D Thermography

64 p.54 Which is the odd one out among these blood disorders?

A Agranulocytosis | B Leukemia | C Mononucleosis | D Septicaemia

65 p.44 Which genetic blood condition is a dominant trait?

A Haemophilia | B Hyperlipidaemia | C Sickle-cell disease | D Thalassaemia

66 p.62 In terms of protein content, which is the odd one out?

A Beans | B Eggs | C Fish | D Meat

67 p.56 What causes elephantiasis, the condition in which a person's leg may swell to an enormous size?

A A bacterium | B A fungus | C A virus | D A worm

68 p.110 What does a blood count measure?

A Sugar and salt content | B Quantity of blood | C How quickly blood clots | D Numbers of blood cells

69 p.50 The muscular tissue responsible for contractions of the intestines is known as?

A Gastric muscle | B Peristaltic muscle | C Smooth muscle | D Sygmoid muscle

70 p.132 What sort of medical device is a Jarvik?

A Artificial heart | B Heart-lung machine | C Replacement hip | D Prosthetic leg

Pot luck
A round of general questions.

71 p.136 What is the name for a doctor's written instructions to a pharmacist to supply medicines?

72 p.64 What nickname was given to British sailors because they were issued with citrus juice to ward off scurvy?

73 p.66 What do the initials MSG stand for?

74 p.100 Claustrophobia is fear of what?

75 p.106 Palpitations are irregular or abnormally fast what?

76 p.122 What part of the body do reflexologists massage?

77 p.120 In the old nursery rhyme 'Jack and Jill', what folk remedy did Jack use to 'mend his head'?

78 p.120 Where is the most famous Roman spa in Britain?

79 p.140 What is the common term for offspring created and selected for particular desired characteristics?

80 p.54 What aquatic creatures were (and sometimes still are) used to cause deliberate bleeding, in an attempt to cure disease?

Short forms
What do the following abbreviations stand for?

81 p.106 ECG

82 p.104 GP

83 p.130 ICU

84 p.98 IQ

85 p.86 IUD

86 p.86 IVF

87 p.90 MMR

88 p.118 MS

89 p.50 RSI

90 p.84 STI

Body sense
Ten questions on different parts of the body.

91 p.70 'Melancholy' literally means 'black choler'. What organ of the body did this originally refer to?

92 p.40 What percentage of the human body consists of water?

93 p.44 What 'c' word, from Greek words meaning colour and body, is the name of the structures in the nucleus of cells that contain DNA?

94 p.76 What is the name of the opening in the middle of the iris in the eye?

95 p.48 The body's biggest and smallest bones are about a metre (3-4 ft) apart in an adult. Where are they?

96 p.50 The biggest (bulkiest) and smallest muscles in the body are about 80 cm (2½ ft) apart. Where are they?

97 p.68 What is the name of the whole of the long tube, extending from the mouth to the anus, where food is digested?

98 p.54 In what tissues of the body are blood cells made?

99 p.84 What is the proper name for the neck of the uterus (womb)?

100 p.78 What kind of sensory nerve signals does the auditory nerve carry?

True or false?
Decide whether the following statements are correct.

101 p.116 Heart failure is the same thing as a heart attack.

102 p.114 Flu viruses mutate and change slightly every year.

103 p.66 A petite woman with a sedentary lifestyle consumes less energy in 24 hours than a 100-watt light bulb.

104 p.72 A single nerve cell can connect with up to around 250 000 others.

105 p.92 High levels of high-density cholesterol in the blood reduces life expectancy.

106 p.128 Before the introduction of antiseptics as many as one in two people undergoing internal surgery died as a result of infection.

107 p.134 The first artificial kidney machine used sausage skins to filter blood.

108 p.132 Ancient Roman dentists performed tooth implants with wrought-iron false teeth.

109 p.114 Rickettsiae live in blood-sucking insects.

110 p.48 Human males are the only primates (apes, monkeys, etc) who do not have a bone in their penis.

For answers and more facts go to the page given below each question number.
For quick answers to complete quizzes 7 to 13 go to page 146.

Fact and fiction

A selection of teasers on the real and the imaginary.

111 In the 1970s Watergate scandal, what was the code name used by the investigating journalists for their informant?
p.60

112 In which Shakespeare play does the Earl of Kent accuse Oswald of being 'A knave…a lily-livered, action-taking knave'?
p.70

113 An annoying loud noise can make you ill. What is the condition called: hypacusia or hypercusis?
p.78

114 Marcel Proust based his famously long novel on the memory evoked by the taste of a madeleine cake dipped in tea. What is the novel's English title?
p.80

115 What is the title of J.D. Salinger's 1951 novel of a 16-year-old expelled from school?
p.90

116 In what fictional Scottish village did Dr Finlay practice?
p.104

117 In which fictional hospital is *Casualty* set?
p.104

118 In which science fiction film are Raquel Welch and fellow medics shrunk to microscopic size?
p.140

119 What branch of science is trying to make such microscopic machines a reality?
p.140

120 What is the name of the type of polymers that are central to current nanotechnology research?
p.140

Prescription please

Match the ailments with the medications in the list below.

121 Anaphylactic shock
p.138

122 Bipolar disorder
p.100

123 Breast cancer
p.116

124 Cold/flu
p.138

125 Cold sore
p.114

126 Heartburn (chronic)
p.138

127 Heart failure
p.138

128 HIV infection
p.114

129 Malaria
p.136

130 Schizophrenia
p.100

Aciclovir
Adrenaline
Anastrazole
Aspirin
Chlorpromazine ('Largactil')
Digoxin
Lithium carbonate
Quinine
Ranitidine ('Zantac')
Zidovudine (AZT)

Little and large

Select the correct option from the four possible answers.

131 The smallest blood vessels are called capillaries; roughly how wide are they?
p.52

A 1 mm — B 0.1 mm
C 0.01 mm — D 0.001 mm

132 About how long is an eight-week-old embryo or foetus?
p.86

A 6 mm (1/4 in) — B 25 mm (1 in)
C 50 mm (2 in) — D 100 mm (4 in)

133 How much blood is pumped by the heart In a year?
p.52

A 1.5 million litres — B 2.7 million litres
C 3.2 million litres — D 4.1 million litres

134 What is the heaviest weight recorded for a human being?
p.40

A 254 kg (560 lb) — B 381 kg (840 lb)
C 444 kg (980 lb) — D 635 kg (1400 lb)

135 What area does an average adult's skin cover?
p.46

A 1.5 m^2 (16 sq ft) — B 2 m^2 (21 sq ft)
C 2.5 m^2 (27 sq ft) — D 3 m^2 (32 sq ft)

136 The skin on the soles of your feet is about how thick?
p.46

A 1 mm — B 2 mm
C 4 mm — D 6mm

137 What volume of air does an adult at rest take in and breathe out in an average breath?
p.58

A 0.5 litres (1/10 gallon) — B 1.2 litres (1/4 gallon)
C 3 litres (3/4 gallon) — D 4.8 litres (1 gallon)

138 Roughly how much food does a person eat in a lifetime?
p.62

A 10 tonnes — B 20 tonnes
C 30 tonnes — D 40 tonnes

139 The outside of the brain's cortex is deeply folded. About what area would it cover if opened out flat?
p.72

A 1000 cm^2 (1 sq ft) — B 1600 cm^2 (1 3/4 sq ft)
C 2400 cm^2 (2 1/2 sq ft) — D 3200 cm^2 (3 sq ft)

140 How much bigger is a human egg than a human sperm?
p.86

A 25 times — B 250 times
C 2500 times — D 25 000 times

Screen symptoms

These answers all have a movie or TV connection.

141 p.100 Which film, starring Jack Nicholson and directed by Milos Forman, is set in a psychiatric hospital?

142 p.104 In which 1960 film does the richest woman in the world (played by Sophia Loren) fall in love with a poor Indian doctor (Peter Sellers)?

143 p.106 What is the name of the song from the same film in which Sophia Loren complains that her 'pulse begins to race'?

144 p.118 Which 1986 television series and 2003 film, written by Dennis Potter, depicts the skin condition psoriasis?

145 p.100 What is the name for a fear of spiders – also the title of a 1990 film?

146 p.80 Name the 1992 film that stars Al Pacino as a blind ex-soldier?

147 p.100 What 1996 Australian film depicts a concert pianist's struggle with schizophrenia?

148 p.44 Frank Sinatra had a recessive genetic trait that led to his nickname. What was the trait?

149 p.100 In *As Good as it Gets*, Jack Nicholson plays a character with which mental disorder?

150 p.124 What 1992 film, starring Nick Nolte and Susan Sarandon, is about a quest for an alternative cure for a little boy's brain disease?

Odd one out

Which one doesn't fit in each of the following lists?

151 p.42 Brain; heart; nerve; kidney

152 p.90 Cold sore; eczema; impetigo; ringworm

153 p.130 Clamp; forceps; speculum; thermometer

154 p.86 GIFT; IVF; IUD; ZIFT

155 p.58 Asbestosis; asthma; emphysema; pleurisy

156 p.126 In terms of disease prevention – dengue fever; HIV; malaria; meningitis

157 p.60 Gingivitis; periodontitis; pulpitis; trench mouth

158 p.48 Osteoarthritis; osteoporosis; Paget's disease; rickets

159 p.116 Erythema; metastasis; pityriasis; rosacea

160 p.138 Penicillin; cyclosporin; tetracycline; streptomycin.

Sensation

These questions are all to do with our senses.

161 p.78 Where in your body would you find the utricle and saccule?

162 p.78 What kind of sensation do the semicircular canals detect?

163 p.78 What kind of sensation, specifically, is measured in hertz?

164 p.78 The auditory ossicles are small bones that transmit sound through the middle ear. What else do they do to it?

165 p.78 How many times louder, in terms of sound intensity, is a 90-decibel train rushing by than a 60-decibel office conversation?

166 p.50 Proprioceptors (literally 'self-receptors') respond to stimuli within the body. In which tissues are they mainly found?

167 p.80 The sense of touch detects at least five sensations other than itching and simple touch. Name any three.

168 p.80 What are the better-known terms for the senses of olfaction and gustation?

169 p.80 Apart from sweet, sour, salt and bitter, a fifth basic taste, called umami, has been identified. What common flavour enhancer tastes of this?

170 p.80 A ruptured eardrum can clearly damage hearing, but what other sense can it affect?

Signs and symbols

Name or explain the following signs and symbols.

171 p.64 Which nationality was the first known to use citrus fruits to counter scurvy at sea?

172 p.110 What is the common name for the polymerase chain reaction, used to generate a big enough sample for DNA 'fingerprinting'?

173 p.114 SARS swept parts of the Far East in 2003. What does SARS stand for?

174 p.122 What is the traditional Indian system of medicine, dating back to ancient holy writings, the *Vedas*?

175 p.126 Against what disease did the Chinese use a form of inoculation, called variolation, some 700 years before doctors in the West?

176 p.126 What public health infrastructure did ancient Mohenjo Daro and Rome have that 1800 London didn't?

177 p.128 What is the 't' word for the practice of boring holes in the skull?

178 p.130 What is the name for the type of surgery used to rejoin severed nerves or blood vessels?

179 p.140 What is the name for the technique of checking test-tube embryos for inherited disorders?

180 p.140 What special type of antibodies could be used to target cancer cells?

**For answers and more facts go to the page given below each question number.
For quick answers to complete quizzes 14 to 18 go to page 146.**

Brother/Sister
Sister/Sister

Brother/Brother
Mother/Son
Mother/Daughter
Brother/Brother

Grandson/Grandfather
Sister/Sister
Father/Son
Son/Father

Family ties

Can you spot the family likenesses between the top and bottom groups?

True or false

Decide whether these statements are correct.

191 p.46 Chimpanzees have two to three times as many body hairs as humans.

192 p.50 Muscles can both pull and push.

193 p.84 Ovulation (egg release) takes place about mid-way between a woman's menstrual periods.

194 p.60 Men have a deeper voice than women because they have a larger voice box.

195 p.76 Colour-blindness is an inability to see any colours.

196 p.62 There is no need to consume fat in your diet.

197 p.142 In Britain today 80 per cent of people live over 60 years.

198 p.88 The heaviest recorded newborn baby weighed over 10 kg (22 lb).

199 p.48 Bone is as strong as iron.

200 p.80 Your tongue has millions of taste buds on its surface.

Illness and cures

A round of questions on diseases and treatments

201 p.142 Which disease was unknown in 1970, but today is one of the highest infectious causes of death worldwide?

202 p.138 People in many warm parts of the world have to take pills regularly to prevent what mosquito-borne disease?

203 p.126 One type of which serious disease, a particular risk to young babies, is prevented by 'Hib' vaccine?

204 p.58 What is the medical name for coalminer's (or coal-worker's) lung?

205 p.100 What is the severe mental disorder that involves hallucinations and delusions of persecution?

206 p.130 What machine, invented in 1960, can be used to repair a detached retina in the eye without actually touching it?

207 p.132 Which is the more common operation in the developed world today: appendix removal or hip replacement?

208 p.138 What do diuretics make you do?

209 p.82 What hormone drug do people with 'Type 1' diabetes have to inject regularly?

210 p.138 What substance, produced in allergic reactions, do antihistamines block?

Male and female

Ten questions on the sexes and their differences.

211 p.90 What is the collective name for features that develop (differently in boys and girls) at puberty, such as breasts and body hair?

212 p.84 What are the main male and female sex hormones that cause the bodily changes of puberty?

213 p.86 An oocyte is an immature what?

214 p.84 What is the main job of the prostate gland?

215 p.84 The clitoris is the female equivalent of which male organ?

216 p.86 What sex is (a) an XX and (b) an XY embryo?

217 p.116 What are the two most common types of cancer of the male reproductive organs?

218 p.92 At about what age do men's testosterone levels start to decline?

219 p.92 Which hormone, usually known by its initials, is responsible for 'hot flushes' during the menopause?

220 p.126 For what two diseases are women in the UK routinely given screening tests?

Around the world

A country-themed round of medical questions.

221 p.64 People in Derbyshire and Switzerland used to get the swollen neck condition called a goitre. What element was lacking from their diet?

222 p.62 Kwashiorkor is the African name for what kind of deficiency disease?

223 p.102 On which Greek island did Hippocrates live?

224 p.102 In which country in the Middle Ages was Europe's first medical school established?

225 p.120 Which world-famous medicinal herb garden, founded in London in 1676, still flourishes?

226 p.120 Which steam-heat treatment may have originated with the Scythians?

227 p.122 The garden of raked gravel in Kyoto, Japan, is a beautiful setting for meditation in what school of Buddhism?

228 p.122 Chinese physicians and acupuncturists believe the life-force flows through special lines in the body. What are these lines called?

229 p.136 What is the English name for *Artemisia*, a herb used for centuries in China to counter malaria?

230 p.142 Place theses countries in order of life expectancy at birth (highest to lowest): (a) Canada; (b) France; (c) Japan; (d) Sweden; (e) UK

For answers and more facts go to the page given below each question number.
For quick answers to complete quizzes 19 to 25 go to pages 146 and 147.

When did it happen?

Identify the decade of these 19th or 20th-century events.

231 The introduction of fluoxetine ('Prozac'). p.136

232 The invention of the stethoscope. p.106

233 The introduction of beta blockers. p.136

234 The introduction of aspirin. p.136

235 The publication of Dr Benjamin Spock's manual, originally called *The Common Sense Book of Child Care*. p.90

236 The introduction of the first transistorised hearing aid. p.78

237 The manufacture of morphine from opium. p.136

238 The first use of laparoscopy – viewing inside the abdomen through a narrow tube. p.108

239 Mass-produced tissue-cultured skin became available for grafting. p.132

240 The discovery of streptomycin. p.136

Pot luck

Test your knowledge with this mixed bag of teasers.

241 Two types of small bodies within a human cell contain DNA. Name them both. p.44

242 What is the body's only bone that is not attached to any other? p.48

243 B cells and T cells are examples of what kind of white blood cell found in the lymphatic system? p.56

244 Name the small, wrinkled part at the very back of the brain that coordinates skilled movements. p.72

245 What substance makes your muscles hurt after strenuous exercise? p.94

246 What are the French scientists Albert Calmette and Camille Guérin famous for? p.126

247 From what Latin word – similar to the modern French term – does the word 'surgery' come? p.128

248 What is the name of the body's 'master cells', able to grow into any type of body tissue? p.140

249 Which Stockholm medical university selects the winners of the Nobel prize for physiology or medicine? p.144

250 What nationality was Bernardo Houssay who, in 1947, became the first South American to win a Nobel prize for medicine? p.144

Who or what?

Select the correct option from the four possible answers.

251 What world record did Iranian weightlifter Hossein Rezazadeh set in 2000? p.94
A 245 kg (540 lb) B 250 kg (551 lb) C 263 kg (580 lb) D 305 kg (672.5 lb)

252 Which is the odd one out among these diets? p.66
A Robert Atkins B William Hay C Susanne Somers D South Beach

253 Which of the following terms best describes the passive-aggressive personality type? p.100
A Argumentative B Impulsive C Indifferent D Weak-willed

254 Prickly heat is caused by? p.118
A An allergy B Blocked sweat glands C Sunburn D Too much alcohol

255 The term for listening to a patient's heart and chest is? p.106
A Auscultation B Percussion C Palpation D Stethoscope

256 What is 'Mitochondrial Eve'? p.44
A A 1960s rock group B A hormonal inbalance C A stage of cell division D A human ancestor

257 The correction of 'buck teeth' is an example of which dental speciality? p.60
A Endodontics B Orthodontics C Periodontics D Prosthodontics

258 Which of these is the odd one out? p.136
A Ernst Chain B Alexander Fleming C Howard Florey D Gerhard Domagk

259 What do tendons do? p.50
A Form tissue in a joint B Join muscles to bones C Link two joints D Transmit pain signals

260 'Beaumont's window' refers to which part of the body? p.68
A The navel, or belly-button B The stomach wall C The eyes D The ears

Know your body

Ten questions on the different parts of the body.

261 p.70 Which organs produce urine?

262 p.70 Which organ stores urine before it is expelled from the body?

263 p.48 The following are all examples of what: hinge; pivot; ball-and-socket; saddle?

264 p.60 Where in your body would you find dentine and enamel?

265 p.84 Enlargement of which part of the body causes a boy's voice to break when he reaches puberty?

266 p.54 To within a litre or a pint, how much blood is in an average adult body?

267 p.54 What common metallic element is found in red blood cells?

268 p.68 It has a capacity of about 1.5 litres (about 2½ pints), it holds swallowed food and it starts the digestion of proteins. What is it?

269 p.58 Name the two vital organs (one single, one pair) that are in your chest cavity.

270 p.52 How many chambers does the heart have?

Mixed bag

Ten questions on a range of topics.

271 p.120 In *Nicholas Nickleby*, what purgative remedy (laxative) were the inmates of Dotheboy's Hall given?

272 p.120 And what is the modern name for the first ingredient of this purge?

273 p.78 The earliest type of hearing aid was often shaped like a saxophone. What was it called?

274 p.92 What is the common name for alopecia?

275 p.40 Iron is a vital component of blood. Which contains more iron: the human body, or six 5 mm (2 in) nails?

276 p.66 Tartrazine and cochineal are examples of what kind of food additive?

277 p.46 Finger and thumb prints are used to identify people. Name two other features (apart from the face) that can be used.

278 p.88 Which parts of a baby are liable to be born first in a 'breech presentation'?

279 p.110 What is the best known type of 'dipstick' medical test, usable at home?

280 p.118 What is the common name for a cerebral haemorrhage or thrombosis?

True or false?

Decide whether these statements are correct.

281 p.40 Neanderthals are a different species from modern humans, but belong to the same genus, Homo.

282 p.90 On average, boys are slower than girls at developing social skills but quicker at developing physical skills such as walking, jumping and hopping.

283 p.96 Cockroaches never sleep.

284 p.92 Only men suffer from the kind of hair loss – known as androgenic alopecia – that happens slowly over years.

285 p.114 HIV destroys some of the cells responsible for defence against disease.

286 p.94 Walking helps to improve aerobic fitness.

287 p.130 A drip is used to drain fluid from the body.

288 p.118 Gout is much more common in men than women.

289 p.96 Humans and dogs are the only creatures that show evidence of dreaming.

290 p.80 A larger area of the brain is devoted to touch sensations from the face than from the whole of the trunk.

Colour me

The answers include at least one colour.

291 p.64 What two liquid food supplements were issued to British children in World War II to boost their intake of vitamins?

292 p.114 What mosquito-borne viral disease is named after the colour of sufferers' skin?

293 p.54 The flag of which organisation honours the blood on the field of the 1853 battle of Solferino?

294 p.76 In the most common form of colour blindness, which two colours are confused?

295 p.84 An ovum, or egg, matures inside a follicle within the ovary. After the mature egg is released, the empty follicle is known as a *corpus luteum*. What colour is it?

296 p.102 Name the four 'humours' that were believed to govern a person's health?

297 p.54 Which type of blood cells transport carbon dioxide from the tissues to the lungs?

298 p.76 What is another name for the fovea, the part of the retina where vision is sharpest?

299 p.46 The skin and hair pigment melanin occurs in dark brown and what other colour?

300 p.108 What kind of rays does thermography detect?

**For answers and more facts go to the page given below each question number.
For quick answers to complete quizzes 26 to 33 go to page 147.**

Latin lesson

Give the common name for these anatomical names.

301 Costa
p.48

302 Femur
p.48

303 Patella
p.48

304 Phalanx
p.48

305 Vertebra
p.48

306 Gastrocnemius
p.50

307 Trachea
p.58

308 Larynx
p.60

309 Duodenum
p.68

310 Oesophagus
p.68

Screen test

Ten questions relating to films and TV.

311 Name either of the actors who played Dr Doug Ross or Dr Elizabeth Corday in the TV series *ER*.
p.130

312 Which TV hospital series was first launched in 1961?
p.104

313 Which British TV comedian, in a famous sketch, complained of giving a pint of blood: 'That's very nearly an armful'?
p.54

314 Which US sitcom was based on *One Foot in the Grave*?
p.92

315 Which film of 2001 explores a famous novelist's struggle with Alzheimer's disease?
p.92

316 In which UK TV comedy-drama series was the title character accused of being a 'lily-livered, caramel-kidneyed, custard-coloured cad'?
p.70

317 What was the full name – including rank and nickname – of the character played by Alan Alda in the TV series *M*A*S*H*, set in the Korean War?
p.104

318 What did the initials *M*A*S*H* stand for?
p.104

319 What numbered army unit was M*A*S*H?
p.104

320 Name the two-headed, three-armed character played by Mark Wing-Davey in the TV version of *The Hitchhiker's Guide to the Galaxy*.
p.132

What's wrong?

Identify the illness described in each question.

321 Which cancer of the lymph nodes, named after a 19th-century English doctor, affects people under 30?
p.56

322 Trachoma is one of the most common causes of blindness worldwide. What part of the eye does it affect?
p.76

323 Gigantism and acromegaly are the result of an excess of which hormone?
p.82

324 What is the general name for severe mental impairment, as caused by Alzheimer's disease and CJD?
p.92

325 Name the disorder in which a person falls asleep suddenly and uncontrollably several times a day.
p.96

326 What is 'strep' short for in the phrase 'strep (meaning sore) throat'?
p.112

327 What is the name for cancer-causing agents such as chemicals, ultraviolet rays and cigarette smoke?
p.116

328 Cerebral thrombosis and cerebral embolism are two types of stroke. What is the third?
p.118

329 A barrel-chested appearance is characteristic of which lung disease?
p.118

330 What phrase is used to describe the immune system's reaction against a transplanted organ?
p.132

And the winner is . . .

Match the achievement with the Nobel prize winner.

331 The discovery of X-rays.
p.144

332 The discovery of the cause of tuberculosis.
p.144

333 The development of the electrocardiogram.
p.144

334 The discovery of the role of chromosomes.
p.144

335 Development of the first sulpha drug.
p.144

336 The discovery of streptomycin as a cure for tuberculosis and meningitis.
p.144

337 The discovery of the structure of insulin.
p.144

338 The development of the CAT scanner.
p.144

339 The development of the technique of DNA amplification.
p.144

340 The discovery of prions as a cause of disease.
p.144

Gerhard Domagk
Willem Einthoven
Godfrey Houndsfield
Robert Koch
Thomas Hunt Morgan
Kary Mullis
Stanley Prusiner
Wilhelm Roentgen
Frederick Sanger
Selman Waksman

C sharp

All the answers begin with the letter C.

341 p.70 What is the medical name for a bladder infection, more common in women than men?

342 p.64 Which metallic mineral, needed for building bones, is found in good quantities in milk and dairy products?

343 p.116 What is the name for the use of anticancer drugs to destroy tumours?

344 p.48 Which relatively soft material, also found in the body's joints, makes up much of a baby's bones?

345 p.54 What (English) 'c' word is sometimes used for blood cells?

346 p.62 The blood levels of what lipoprotein are raised by a diet rich in saturated fats?

347 p.138 What complaint are laxatives used to relieve?

348 p.128 What is the common name for the clear liquid with a pungent smell – chemical formula $CHCl_3$ – once used as an anaesthetic?

349 p.90 What is the name of the thick, scaly rash that often appears on a young baby's scalp?

350 p.92 Which disease (or group of diseases) is/are the greatest single health risk for women in their 50s and 60s?

Medications

This round is all about cures and treatments.

351 p.100 Tranquillisers and antidepressants are types of drugs used to treat what broad class of illness?

352 p.140 What name, taken from military technology, is applied to treatments that accurately target disease cells?

353 p.136 Which important hormone drug was the first drug to be genetically engineered?

354 p.122 A well-known Asian herbal 'tonic' comes from plants of the genus *Panax*, whose roots supposedly look like a human body. What is the herb called?

355 p.138 What disease, a major cause of death in women, is treated by tamoxifen and anastrazole?

356 p.136 What name is given to a book of plant medicines?

357 p.136 Which plant does aspirin come from?

358 p.136 Which plant does morphine come from?

359 p.136 Which plant does quinine come from?

360 p.136 And which plant does atropine come from?

Chain reaction

All these questions concern the chemistry of the body.

361 p.62 What three chemical elements are found (in different proportions) in both fat and carbohydrate molecules?

362 p.62 What well-known type of food chemical is a complex carbohydrate?

363 p.62 What are the common names of lactose and fructose?

364 p.68 What is the name for the chemical substances (proteins) that break food molecules into smaller, simpler substances?

365 p.68 What are the simple substances that result from the digestion of carbohydrates?

366 p.68 What are the simple substances (two types) that result from the digestion of fats?

367 p.70 One of the jobs of the liver is to render chemicals, including poisons, harmless. What one-word term describes this process?

368 p.70 The liver converts ammonia, a breakdown product of proteins, into what harmless substance that is then excreted in urine?

369 p.80 What makes ethyl mercaptan notorious?

370 p.82 Which gland, found in the neck, regulates the speed at which the body 'burns' energy?

Whodunnit?

Can you identify the following people?

371 p.144 British molecular biologist who is the only person to have won two Nobel prizes for chemistry (1958 and 1980), both for work on the structure of important biological molecules.

372 p.126 French scientist who first developed vaccines against anthrax and rabies.

373 p.98 French philosopher who believed the pineal gland to be the meeting point of mind and body.

374 p.52 South African doctor who performed the first successful human heart transplant.

375 p.94 Bestselling author and advocate of running for fitness who died of a heart attack when out on his daily run in New York in 1984.

376 p.52 English doctor who discovered the circulation of blood.

377 p.96 Austrian doctor who in 1890 put forward an influential theory of the meaning of dreams.

378 p.98 Russian physiologist who discovered the conditioned reflex by experiments on dogs.

379 p.102 British researcher who discovered penicillin.

380 p.128 British surgeon who introduced the use of carbolic acid as an antiseptic in operating theatres.

**For answers and more facts go to the page given below each question number.
For quick answers to complete quizzes 34 to 40 go to pages 147 and 148.**

True, false or maybe?

Decide whether the following statements are correct.

381 p.94 Exercise always helps you to lose weight.

382 p.50 Cardiac muscle can contract of its own accord, without being stimulated by nerve signals.

383 p.136 Quinine was introduced to Europe by Jesuit missionaries.

384 p.86 Identical twins result from a fertilised egg splitting in two.

385 p.86 Female babies are born without eggs – these are made by the ovaries at puberty.

386 p.46 People of black Afro-Caribbean origin have ten times as many pigment-producing melanocyte cells as 'white' people.

387 p.90 By the age of 14 boys are, on average, taller and heavier than girls.

388 p.142 Infant mortality in the UK in 1900 was 20 times higher than it is today.

389 p.124 Homeopathic practitioners believe that the more pure and concentrated a remedy is, the more powerfully it acts.

390 p.86 The placenta develops from the lining of the uterus (womb) to nourish a baby.

Small-scale

A set of brainteasers on the smaller things in life.

391 p.42 Where would you find a nucleus, Golgi aparatus, ribosomes and mitochondria?

392 p.42 Organs are to the human body as organelles are to what?

393 p.56 What is the name of giant scavenger cells found in lymph nodes?

394 p.78 Name the three smallest bones in the human body.

395 p.110 What type of microorganisms are grown on a culture medium for examination under a microscope?

396 p.112 What is the medical name for microorganisms and other agents that cause disease?

397 p.64 What are the two types of micronutrients that form small but vital parts of food?

398 p.62 Does a gram of fat give more, less or about the same amount of energy as a gram of carbohydrate?

399 p.140 Gene therapy aims to replace defective genes with normal ones. What type of microorganism could be used as a 'carrier' for replacement genes?

400 p.140 What is the name given to a hybrid cell made from a white blood (B) cell and tumour cell, and used to produce monoclonal antibodies.

How many?

Select the correct option from the four possible answers.

401 p.42 Approximately how many cells are there in an adult body?
- A 10 million
- B 10 billion
- C 10 trillion
- D 10 quadrillion

402 p.46 Roughly how many sweat glands are there on your body?
- A 100 000
- B 500 000
- C 1 million
- D 2.5 million

403 p.60 How many milk teeth are in a full set of child's 'milk' teeth?
- A 16
- B 20
- C 24
- D 28

404 p.48 How many bones are there in the adult human?
- A 206
- B 216
- C 226
- D 236

405 p.54 A 1 mm^3 (1 microlitre) drop of blood contains approximately how many red blood cells?
- A 1 million
- B 5 million
- C 10 million
- D 15 million

406 p.72 Approximately how many nerve cells are there in the brain?
- A 100 million
- B 1 billion
- C 10 billion
- D 100 billion

407 p.72 Up to how many brain cells die, approximately, every day?
- A 1000
- B 10 000
- C 100 000
- D 1 million

408 p.86 On average, how many sperm does a fertile man ejaculate at one time?
- A 1 million
- B 10 million
- C 100 million
- D 250 million

409 p.86 About how many immature eggs are in a woman's ovaries?
- A 2000
- B 20 000
- C 200 000
- D 2 million

410 p.142 About 2500 babies died in Britain in 1995; what was the equivalent figure for 1949?
- A 10 000
- B 16 000
- C 22 000
- D 27 000

Songs and books

The answers all relate to parts of the anatomy.

411 p.52 In the song, what part of the anatomy was left in San Francisco?

412 p.40 Fats Waller complained that 'your pedal extremities are colossal'. What part of the body did he mean?

413 p.70 What part of the body did Dr Lecter eat with 'fava beans and a nice chianti' in *Silence of the Lambs*?

414 p.46 Which band had 'Perfect Skin' in the 1980s?

415 p.76 Singers including Dick Powell, The Flamingoes and Art Garfunkel 'don't know if it's cloudy or bright'. Why?

416 p.52 Who wrote *The Heart of Darkness*?

417 p.48 The Old Testament book of Ezekiel includes an anatomy lesson, turned into a spiritual with the repeated line: 'I hear the word of the Lord'. What is the song called?

418 p.52 What went 'Zing!' when you smiled at me, according to singer Ruth Etting in 1935?

419 p.76 What was the name of the 'pretty little black-eyed girl' in a hit song by Elton John?

420 p.70 In which series of children's books does Captain Haddock use a vast range of curses, including 'lily-livered bandicoots'?

Mind your Ps

In this round, the answers all start with the letter P.

421 p.58 What medical term describes anything to do with the lungs?

422 p.84 Name the chemical substances involved in sexual attraction in many creatures.

423 p.100 What is the proper term for 'talking cures' used for some kinds of mental disorders?

424 p.106 What is the medical term for feeling and pressing on the abdomen to check for swellings or lumps?

425 p.112 What is the name of the abnormally shaped protein molecules believed to cause BSE and variant CJD?

426 p.118 The 1994 film *The Madness of King George* showed George III's descent into mental confusion. What often inherited disorder is believed to have caused his madness?

427 p.64 Bananas are a rich source of which mineral?

428 p.126 The UN has declared smallpox eradicated. Which disease does it next hope to eliminate by vaccination?

429 p.120 What breakfast dish is supposed to help acne if applied to the skin?

430 p.142 Which book of the Bible declares that 'The days of our age are three-score years and ten'?

When did it happen?

Ten questions about dates and times.

431 p.76 In which decade did the first plastic contact lenses become available?

432 p.126 Which was discovered first: the vaccination for polio or for rabies?

433 p.126 When (to within two years) was the last recorded case of smallpox?

434 p.102 In which century did the Black Death sweep Europe and parts of Asia?

435 p.144 Since World War II, US researchers have dominated the Nobel prizes in medicine. When (to within three years) did an American first win this prize?

436 p.102 When (to within one year) was HIV, the virus that causes AIDS, discovered?

437 p.128 When did Leonardo da Vinci make his famous anatomical drawings?

438 p.132 When (to within two years) were the first heart pacemakers fitted?

439 p.104 In which century was Britain's Royal College of Physicians founded?

440 p.128 By the 10th century, surgeons in Moorish Spain were performing what kind of eye operation?

Medical technology

Test your knowledge of doctors' aids and equipment.

441 p.60 What medical specialist uses a drill, probe, mirror and X-ray machine?

442 p.106 In a blood-pressure reading (say, 120/80), the second figure is the diastolic pressure. What is the first?

443 p.108 The Doppler effect explains the changing pitch of a police-car siren as it passes. What bodily function is studied using the same effect?

444 p.108 What is the name of the technique of 'mapping' the body's surface temperature?

445 p.108 What is the term for examining inside an organ such as the stomach through a narrow flexible tube?

446 p.144 What method of medical imaging led to Nobel prizes in both 1952 and 2003?

447 p.108 Echocardiography is a type of ultrasound scanning used to study which organ?

448 p.132 A cochlear implant is used to help people, especially children, with what disability?

449 p.134 Apart from artificially oxygenating and circulating the blood, what other important blood function does a heart-lung machine perform?

450 p.134 Name the machine used to pulverise kidney stones with ultrasound.

For answers and more facts go to the page given below each question number.
For quick answers to complete quizzes 41 to 45 go to page 148.

QUIZ 45 CHALLENGE

He cells, she cells...

Match the names of human cells with their pictures below.

- White blood cell
- Red blood cell
- Liver cell (hepatocyte)
- Skin cell
- Nerve cell
- Cardiac muscle cell
- Epithelial cells
- Fat cell
- Sperm cell
- Ovum cell

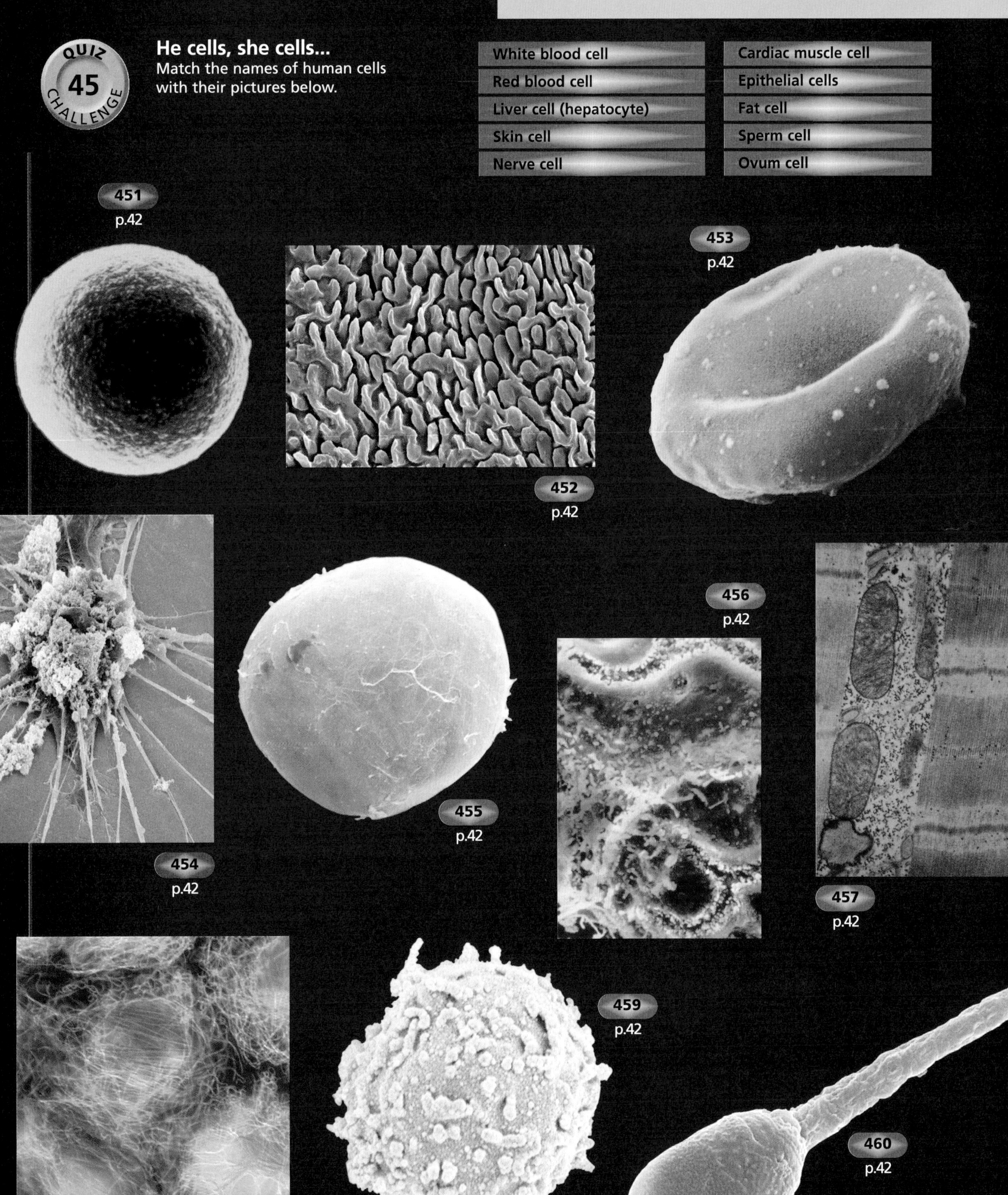

What's wrong?

A set of questions on medical conditions.

461 Inability to sleep is known as what?
p.96

462 What common virus disease can cause malformation of a foetus if a pregnant woman catches it?
p.86

463 People with what disorder often use a home analyser to test a pin-prick drop of blood?
p.110

464 What are the two most important types of germs?
p.112

465 Of what physical condition are the following all examples: comminuted, compression, greenstick and transverse?
p.48

466 What is the proper name for the 'glands' in the neck, armpit and groin that sometimes swell and become tender during illness?
p.56

467 Fluid retention and protein in the urine are two main symptoms of pre-eclampsia (or toxaemia) in pregnant women. What is the third?
p.86

468 What is the medical name for abnormal sadness and despair?
p.100

469 A fever is defined as a temperature of more than what?
p.106

470 The names Sabin and Salk are associated with vaccines against which disease?
p.126

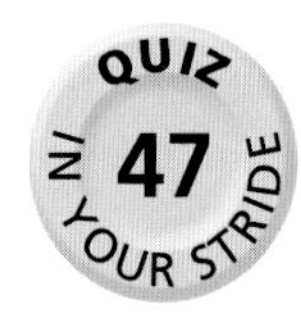

Pot luck

A general round of medical questions.

471 The best-known type of 'smear test' is an examination of cells from where?
p.110

472 Coughing, sneezing and hiccups are all examples of what kind of nervous action?
p.58

473 What vital fluid has no nutritional value yet is an essential part of the diet?
p.62

474 What does the 'E' in E numbers stand for?
p.66

475 What is the name of the condition in which abnormally high pressure builds up in one or both eyeballs?
p.76

476 What is the oldest method of looking inside the body without an operation?
p.108

477 What type of cancer, second most common in the UK to breast cancer, is the biggest cause of cancer deaths?
p.116

478 What form of yoga-based meditation technique did members of the Beatles study under Maharishi Mahesh Yogi?
p.122

479 Which range of flower-based 'remedies' shares a name with a famous composer?
p.124

480 What is the everyday word for pharmaceuticals?
p.136

Fame!

Questions on people who have made the news.

481 Why did Aaliyah Hart make news in 2003?
p.88

482 In the 19th century the famous twins Eng and Chang toured exhibitions. Where were they from?
p.86

483 Which world-renowned physicist has lived for many years with motor neuron disease?
p.118

484 What was the name of the British mother who gave birth in 2003 to a test-tube baby selected as a suitable bone-marrow donor for his older brother?
p.140

485 Which world record did Paul Tergat break in September 2003?
p.94

486 Which Roman statesman supposedly gave his name to surgical childbirth (both names, please)?
p.88

487 Which former bodybuilder became governor of California in 2003?
p.94

488 Which queen pioneered anaesthesia in childbirth, when she was given chloroform?
p.88

489 Who founded the profession of nursing as we know it, and was commemorated on a banknote in the 1980s?
p.102

490 What was the Royal Navy's nickname for the tots of rum (or grog) issued to seamen until 1970?
p.54

True or false?

Decide whether the following statements are correct.

491 Over 1000 years ago Indian surgeons performed plastic surgery to reconstruct a human nose.
p.128

492 Heart-valve transplants are frequently performed using valves from pigs' hearts.
p.132

493 Geneticists believe that the natural age limit for most people is over 130.
p.142

494 Viruses invade or inject their genetic material into living cells and use the cells' materials to reproduce themselves.
p.114

495 Unlike other scanning methods, MRI, thermography and ultrasound have no known risk of long-term ill-effects.
p.108

496 Adults need an average of 10 hours' sleep a night.
p.96

497 An aneurysm is a balloon-like swelling in an artery.
p.52

498 Many, if not most, cases of stomach ulcers are caused by stress.
p.68

499 Stem cells may offer a future cure for Parkinson's disease and diabetes.
p.140

500 The amount of light that passes to the retina in the eye decreases with age.
p.76

**For answers and more facts go to the page given below each question number.
For quick answers to complete quizzes 46 to 52 go to page 148.**

Go with the flow

These answers all have a connection with fluids.

501 p.62 What food featured heavily in the Japanese film *Tampopo*?

502 p.68 The body makes five kinds of digestive juices, including saliva, pancreatic juice and intestinal juice. Name the other two.

503 p.70 Some of the products of the liver's activities go into bile. Where do the rest of them go?

504 p.70 Name the tubes that (a) lead into the bladder, and (b) leads out of the bladder?

505 p.72 What name is given to the spaces inside the brain that contain cerebrospinal fluid?

506 p.76 Name the liquids or 'humours' that fill (a) the body of the eyeball, and (b) the part in front of the lens.

507 p.82 The secretions of exocrine glands are expelled from the body. Where do endocrine secretions go?

508 p.82 What is the name of the small bodies in the pancreas where insulin is made?

509 p.98 Rorschach tests ask subjects to respond to what?

510 p.110 What body fluid did medieval doctors believe to be the most revealing of disease?

Anagram corner

Unravel the anagrams with the help of the clues.

511 p.46 CYAN OMELETS They make dark pigment.

512 p.46 CHORAL FILLIES Beards grow from them (two words).

513 p.58 ANDROID ICEBOX What you breathe out (two words).

514 p.66 MIMOSA BELT All the chemical reactions in the body.

515 p.68 SILENT AILMENTS Duodenum, jejunum plus ileum (two words).

516 p.68 RESILIENT AGENT Absorbs water from digested food (two words).

517 p.72 MERGE RATTY What you think with (two words).

518 p.78 AUTHENTIC SEA SUB Joins the ear and throat (two words).

519 p.80 BETA STUDS All over your tongue (two words).

520 p.82 CLARENDON DESIGN They all produce hormones (two words).

What does it mean?

Select the correct option from the four possible answers.

521 p.42 The cells of the human body (and other living things) are named after?

A Monks' cells	B Electrical cells
C Cockle shells	D Cellulose bodies

522 p.122 Yoga is the Sanskrit word for?

A Control	B Exercise
C Meditation	D Union

523 p.50 Actin and myosin are?

A Amphetamines	B Toxins
C Antibiotics	D Types of muscle filament

524 p.76 A chiasma is?

A A blood clot	B An opthalmic instrument
C A crossing or intersection	D A noxious smell

525 p.72 What is the name for a chemical messenger in the brain?

A A neurotransmitter	B A neuron
C A motor neuron	D A brainstem

526 p.68 A sphincter is?

A A hearing disorder	B A psychological crisis
C A ring of muscle	D A small 'knot' of nerves

527 p.40 What does 'ectomorph' mean?

A A narrow tube	B An outer layer of skin
C A kind of skin graft	D A thin, gangly person

528 p.132 Bioglass is?

A A measuring container	B A type of lens
C An implant material	D A type of microscope

529 p.72 A gap between two nerve cells is called?

A A synapse	B A reflex
C A ganglion	D An antigen

530 p.122 The China Syndrome is?

A A brittle bone disease	B A film with Jane Fonda
C Another term for SARS	D An eating disorder

Body action

Ten questions about how the body works.

531 p.48 Bones contain almost all of the body's supply of which mineral?

532 p.62 What is the proper name for roughage?

533 p.64 Red meat, liver and some fish are among the best sources of which essential metallic mineral?

534 p.68 Where in the body is insulin produced?

535 p.70 What is the medical term for getting rid of body wastes?

536 p.70 To which pair of organs does the adjective 'renal' refer?

537 p.70 Which organ is sometimes referred to as 'the body's chemical plant'?

538 p.84 What is the name of the tubes that lead from the ovaries to the uterus (womb)?

539 p.106 What function is tested by tapping a patient's knee, or scratching the sole of the foot?

540 p.110 One of the oldest immune-system tests consists of observing the clumping together of blood cells. What does the test reveal?

Prescription check

Match the ailments with the medications listed below.

541 p.48 Arthritis

542 p.58 Asthma

543 p.116 Cancer

544 p.62 Constipation

545 p.46 Eczema

546 p.56 Hay fever

547 p.116 Angina

548 p.52 Hypertension (high blood pressure)

549 p.96 Insomnia

550 p.112 Thrush

ACE inhibitor
Antifungal
Antihistamine
Antipruritic
Bronchodilator
Cytotoxic drug
Hypnotic
Laxative
Nitrates
NSAID

Initial letters

What do the following abbreviations stand for?

551 p.100 ADD (a type of children's behaviour)

552 p.40 BMI (referring to weight and build)

553 p.132 BMT (a treatment – for leukaemia, for example)

554 p.108 CAT (a type of body scanning)

555 p.118 COPD (a lung disease)

556 p.72 CNS (neurology)

557 p.100 ECT (a treatment for depression)

558 p.96 EEG (a means of studying brain waves)

559 p.108 MRI (a type of body scanning)

560 p.74 TIA (a mini-stroke)

All Greek to me!

These questions all involve Latin or ancient Greek.

561 p.102 What common term is based on the Latin word for health and cleanliness?

562 p.102 What Latin term is used to refer to a cure-all?

563 p.98 Which study of human processes gets its name from the Greek words for 'mind' or 'soul' and for 'study'?

564 p.98 Which type of behaviour therapy gets its name from the Latin word for 'to learn' or 'to know'?

565 p.90 What is the Latin name for the fine, downy hair on a newborn baby's skin?

566 p.90 And what is the term for the greasy coating on newborn babies (especially premature babies)?

567 p.98 What Latin term did Freud use to refer to the part of the psyche that gives rise to our unconscious or instinctive impulses?

568 p.98 And what term did he use for the conscience?

569 p.50 Where would you find the muscle known as *biceps brachii*?

570 p.50 And where would you find the muscle known as the *gluteus maximus*?

**For answers and more facts go to the page given below each question number.
For quick answers to complete quizzes 53 to 57 go to page 148.**

QUIZ 57 WARM UP

Tools of the trade

Match the medical instruments to the names in the list below.

- Clamping forceps
- Dissecting forceps
- Ear thermometer
- Foetal stethoscope
- Laryngoscope
- Opthalmoscope
- Scalpel
- Sphygmomanometer
- Stethoscope
- Syringe

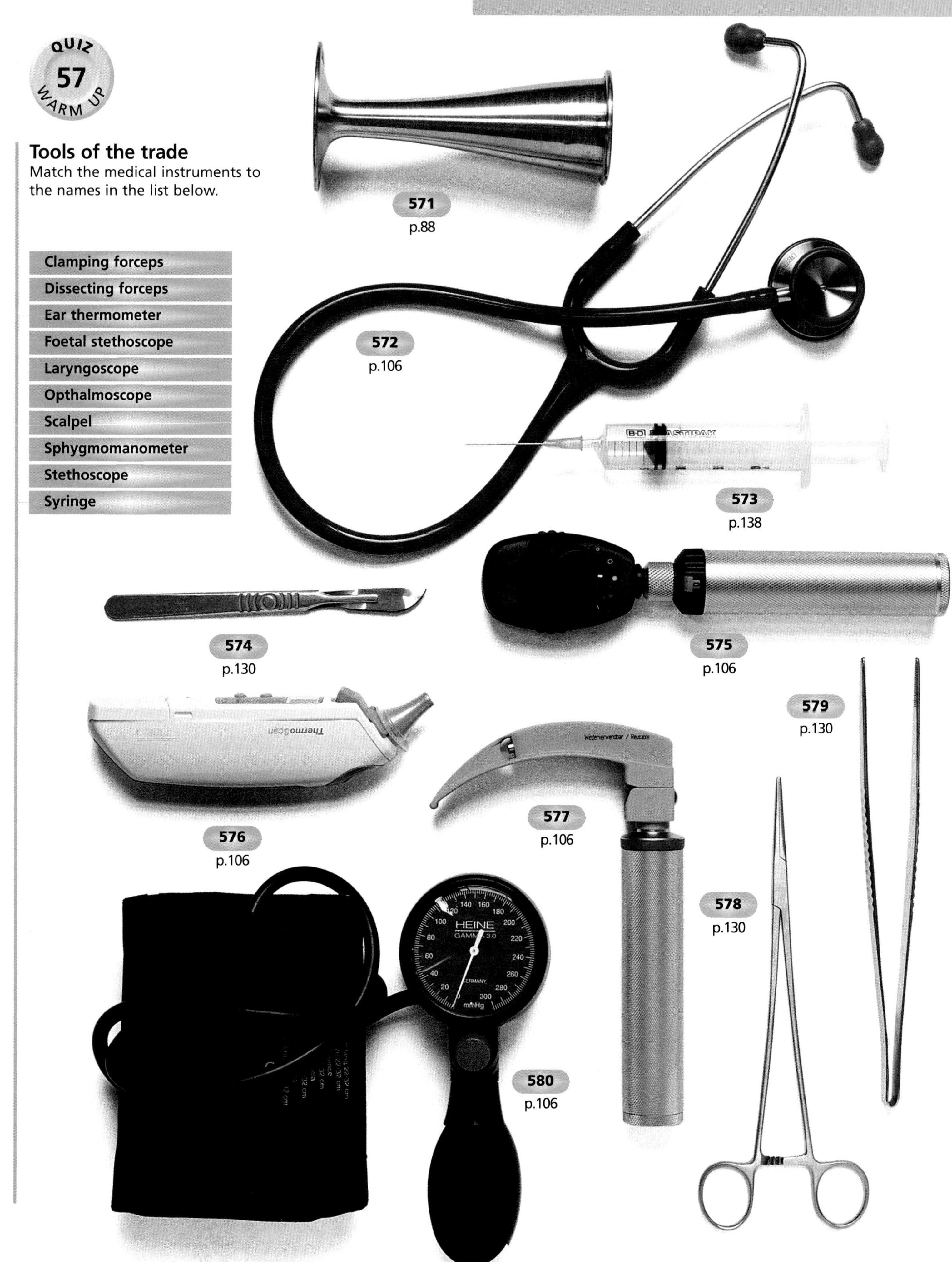

Ills and remedies

Ten questions relating to diseases and treatments.

581 p.120 What strongly flavoured plant, a favourite of herbalists, can lower blood pressure and blood cholesterol levels?

582 p.110 What is the name for a sample of mucus taken from the throat with a cotton-wool bud, to test for infection?

583 p.112 What did the word 'contagious' originally and literally mean?

584 p.112 What group of living things includes the cause of athlete's foot?

585 p.116 What is the medical term for a solid lump, either cancerous or benign?

586 p.116 In angioplasty, what is fed via a catheter into a coronary artery to widen it and relieve angina?

587 p.124 What alternative method claims to harness the 'positive energy' or 'vibrations' in semiprecious stones.

588 p.124 Which renowned psychiatrist at first used hypnosis to explore unconscious thoughts but soon abandoned it?

589 p.134 What is the medical term for the use of X-rays to cure cancer?

590 p.142 Most doctors and lawyers today take the cessation of which organ's function to signal death?

Past and present

A quiz round on medical practices through the ages.

591 p.128 What was the standard method of cauterising battle wounds in the Middle Ages?

592 p.102 What oath, believed to date from ancient Greece, formed the basis of ethical medical conduct?

593 p.110 What is the general name for medical tests that are used to check people for problems when symptoms are not present?

594 p.138 Where, until 1970, might a doctor have written references to minims, grains, scruples and drachms?

595 p.104 In the past, what part of the body would an itinerant occulist have treated?

596 p.120 What is the name for the traditional therapy in which small hollow heated vessels were applied to the skin to draw out harmful 'humours'?

597 p.88 What is the name for a doctor who specialises in childbirth?

598 p.120 In folk medicine, what were cobwebs sometimes used for?

599 p.102 Doctors still use the symbol of Asclepius, Greek god of healing. What creature forms part of the symbol?

600 p.128 What do the red and white stripes on a barber's pole symbolise?

Young and old

Ten questions about growing up and growing old.

601 p.92 What was the authenticated age of the oldest known person when she died?

602 p.96 In which fairy tale is the heroine woken from a long sleep by the kiss of a handsome prince?

603 p.142 What process for preserving bodies was featured in the Woodie Allen movie *Sleeper*?

604 p.90 Who is the narrator in *To Kill a Mockingbird*, the growing-up novel by Harper Lee?

605 p.92 Which film starred Bette Davis and Joan Crawford as two faded Hollywood stars?

606 p.92 Which famous male and female actors starred in the 1981 movie about advancing years, *On Golden Pond*?

607 p.90 Who played the title role in the growing-up movie *Billy Elliot*?

608 p.92 Which Hollywood star, renowned for her eight or nine marriages, married her latest and longest-lasting husband in her late 60s?

609 p.90 Proportionately, a baby grows fastest in its first year. In that year, by how much does its bodyweight increase, on average?

610 p.92 Which branch of medicine deals with the problems of older people?

Action!

A quiz round mainly to do with muscles and movement.

611 p.122 Which Chinese system of movement combines exercise and meditation?

612 p.60 What is the tongue-stretching Maori war chant called?

613 p.52 An artificial pacemaker is an electronic device to regulate what?

614 p.54 What mythical beings are said to creep from their graves to suck the blood of sleeping people?

615 p.74 What is the term for a true knee-jerk reaction?

616 p.94 Starting with the best, put these exercises in order of value for developing muscular strength: aerobics; rowing; tennis.

617 p.94 Starting with the best, put these exercises in order of value for developing flexibility: golf; jazz dancing; squash.

618 p.96 Apart from limb-twitching, what is the most common visible sign that someone is dreaming?

619 p.88 Oxytocin is sometimes given to induce labour at childbirth. What does it do?

620 p.122 People with travel sickness are sometimes advised to press on a point just above the wrist. On what system of healing is this based?

**For answers and more facts go to the page given below each question number.
For quick answers to complete quizzes 58 to 64 go to pages 148 and149.**

A is for . . .

All the answers start with the letter A.

621 p.114 What do the initials AIDS stand for?

622 p.66 Catabolism is one basic aspect of body chemistry. What is the other?

623 p.70 Which waste product does the liver convert into urea?

624 p.82 What glands would you find just above the kidneys?

625 p.88 What is the proper medical name for the 'waters' that precede the birth of a baby?

626 p.110 An ELISA (enzyme-linked immuno-sorbent assay) test on blood plasma is used to detect what indicators of various diseases?

627 p.52 What is the general name for an abnormal or irregular heart rhythm?

628 p.124 The Alexander technique aims to improve posture. What was the profession of its originator?

629 p.130 During surgery, who is responsible for keeping the patient both unconscious and alive?

630 p.138 To what group of drugs do heparin and warfarin belong?

Special subjects

Identify the specialist from the sample activity given.

631 p.104 Which type of surgeon sets a broken leg?

632 p.102 Which specialist studies the spread of AIDS in Africa?

633 p.102 Which specialist monitors the growth of a brain tumour?

634 p.104 Which type of surgeon performs prostate operations?

635 p.102 Which specialist studies the effects of aspirin on the body?

636 p.104 Which specialist treats anaemia?

637 p.102 Which specialist studies the AIDS virus?

638 p.102 Which specialist studies processes in the body, such as the release of energy by the breakdown of glucose molecules?

639 p.104 Which specialist treats problems with the hormone system?

640 p.104 Which type of surgeon performs ear, nose and throat operations?

The dating game

When did the following events take place?

641 p.122 When did Ayurvedic medicine originate in India?
A c.2500 BC B c.1500 BC
C c.1000 BC D c.100 BC

642 p.122 When is acupuncture believed to have originated?
A c.2500 BC B c.1500 BC
C c.1000 BC D c.100 BC

643 p.128 In which year was the first successful surgical removal of an appendix performed?
A 1675 B 1736
C 1788 D 1815

644 p.128 When was cocaine first used as a local anaesthetic?
A 1736 B 1788
C 1815 D 1884

645 p.134 In which year was X-ray radiotherapy first used for cancer?
A 1898 B 1903
C 1908 D 1913

646 p.134 When was the first successful iron lung invented?
A 1927 B 1935
C 1943 D 1951

647 p.44 In which year did Francis Crick and James Watson discover the structure of DNA?
A 1951 B 1952
C 1953 D 1954

648 p.134 When was home kidney dialysis introduced?
A 1954 B 1959
C 1964 D 1969

649 p.52 When was the first human heart transplant performed?
A 1965 B 1967
C 1969 D 1971

650 p.44 When was the human genome published?
A 2000 B 2001
C 2002 D 2003

Pick and mix

A mixed selection of body and health teasers.

651 p.90 What is the name of the hormonal and other changes that herald adolescence?

652 p.100 What is the medical or psychiatric term for an irrational fear of an object or situation?

653 p.118 What is the main cause today of the chest conditions, chronic bronchitis and emphysema?

654 p.132 Which whole organ was the first to be successfully transplanted?

655 p.138 Bronchodilators help with breathing. Which body part do they act upon?

656 p.124 What common ailment is flotation therapy said to relieve?

657 p.70 Inflammation of the liver, often caused by a virus infection, is known medically as what?

658 p.78 What are you equalising when you yawn and your ears 'pop'?

659 p.98 What term refers to mental actions and processes that we are not aware of?

660 p.142 Which two parts of the body – not strictly living tissues – continue to grow for a while after death?

Tricky treatments

Unravel these anagrams of different treatments.

661 p.54 RUSSIAN FONT Lifesaving blood donation.

662 p.96 FIND DRUG MUSE He founded psychoanalysis (two words).

663 p.52 STAR LANTERN PATH Christiaan Barnard did it first (two words).

664 p.108 SCANDALOUS TURN Used to observe unborn baby (two words).

665 p.124 TAHOE TOPSY Manipulative treatment.

666 p.130 CAN HESITATE Banishes pain.

667 p.130 GREY HOKEY RULES Common name for minimally invasive operation (two words).

668 p.132 HELPMATE PRINCE The commonest implant operation (two words).

669 p.136 AIN'T NERVOUS Into a vein.

670 p.142 TEMP MOTORS Happens in the morgue (two words).

Around the body

Test your knowledge of workings of the body.

671 p.56 What is the name of the body system concerned with fighting disease?

672 p.56 Apart from the blood system, another network of tubes carries fluid through the tissues. What is it called?

673 p.60 What 'u' word is the small soft projection hanging down from the soft palate at the back of the mouth?

674 p.60 What 'p' word is the proper medical name for the throat?

675 p.68 Villi greatly increase the internal area of the small intestine. What are they?

676 p.70 Which organ of the body does the adjective 'hepatic' refer to?

677 p.70 Nephritis is a general term for any inflammation or infection of which part of the body?

678 p.72 What is the term for the crinkly outer layer of the brain?

679 p.82 What is the body's so-called 'master' gland, which controls the others?

680 p.110 What body fluid is withdrawn for testing by means of a lumbar puncture?

True or false?

Decide whether the following statements are correct.

681 p.96 Sleep is important for growth and the renewal of bodily tissues.

682 p.96 Most people do not dream every night.

683 p.116 Cancers produce chemicals that stimulate blood vessels to grow and feed them.

684 p.112 Viruses breed rapidly in warm polluted water.

685 p.48 A baby is born with almost 100 more bones than a normal adult.

686 p.92 Many body cells are believed to be 'programmed' to die after dividing a certain number of times.

687 p.68 A 'bilious attack' is caused by bile going the wrong way into the stomach.

688 p.72 'Grey matter' in the brain is caused by the colour of the myelin sheaths that cover nerve cells.

689 p.80 You sense the smell of some chemicals, such as chloroform, with your tongue rather than your nose.

690 p.122 Hundreds of years ago, Chinese herbalists prescribed a plant containing an ingredient now used to treat asthma.

**For answers and more facts go to the page given below each question number.
For quick answers to complete quizzes 65 to 71 go to page 149.**

Stage and screen

A round of questions about films and plays.

691 (p.144) What event is the setting for the 1963 spy thriller *The Prize*, starring Paul Newman?

692 (p.50) What cartoon character used to eat spinach to strengthen his muscles?

693 (p.80) Which blind stage king declared, 'Might I but live to see thee again in my touch, I'd say I have eyes again!'

694 (p.98) Name the British director of the 1980 film *Bad Timing*, the story of an affair between a psychoanalyst and his protégé.

695 (p.66) Which blonde film star starred in the 1933 David Selznick movie *Dinner at Eight*?

696 (p.48) In which Shakespeare play does the main character, holding a skull, proclaim: 'Alas, poor Yorick. I knew him, Horatio; a fellow of infinite jest, of most excellent fancy.'

697 (p.80) Which British actress starred in the 1961 movie *A Taste of Honey*?

698 (p.76) Who played the title role in the original 1933 film *The Invisible Man*?

699 (p.46) Who starred alongside George Raft and Ann Dvorak in the title role of the 1932 gangster movie *Scarface*?

700 (p.60) Which British actress starred in the film *Little Voice*?

Also known as...

Find the medical term – or the more common name.

701 (p.118) What is the more common name for amyotrophic lateral sclerosis, or ALS?

702 (p.78) What is the medical name, reminiscent of orchestral drums, for the eardrum?

703 (p.78) The inner ear has so many complex channels that it is sometimes called what?

704 (p.82) What is another, simpler name for the body's endocrine system?

705 (p.98) What is the alternative, two-word term for psychopathology?

706 (p.98) The German word *Gestalt* is used for a school of psychology. What does it mean?

707 (p.124) What form of therapy, working on the principle of 'helping the body to help itself', was originally called 'nature cure'?

708 (p.138) What is the common name for hypnotics?

709 (p.138) What is the medical name for 'clot-busting' drugs?

710 (p.138) What is the name for a book listing drugs that a doctor may prescribe? There are two alternatives.

A bigger choice

Select the correct option from the four possible answers.

711 (p.56) Which of the following are *not* types of white blood cells?
- A Basophils
- B Antigens
- C Lymphocytes
- D Neutrophils

712 (p.40) Which of the following statements about humans is *not* true? That they are:
- A Invertebrates
- B Animals
- C Mammals
- D Primates

713 (p.62) Which of these is a good source of monounsaturated fat?
- A Butter
- B Lard
- C Olive oil
- D Smoked salmon

714 (p.108) Which of the following medical imaging techniques uses radioactive isotopes?
- A CAT scan
- B MRI
- C PET scan
- D CT scan

715 (p.74) Which of these form part of the peripheral nervous system?
- A The brain
- B Cranial nerves
- C The spinal cord
- D The cerebrum

716 (p.64) What is a better known name for 'accessory food factors'?
- A Vitamins
- B Proteins
- C Carbohydrates
- D Fats

717 (p.56) A single lymphocyte can produce how many antibodies in an hour?
- A 10 000
- B 500 000
- C 1 million
- D 1 billion

718 (p.50) Which of the following is a shoulder muscle?
- A The sartorius
- B The deltoid
- C The peroneus longus
- D The tibialis anterior

719 (p.114) Rabies mainly affects which part of the body?
- A The skin
- B The digestive system
- C The immune system
- D The nervous system

720 (p.52) Which of the following is *not* infectious (caused by germs)?
- A Measles
- B Typhus
- C Dengue
- D Coronary thrombosis

Medical terminology

Test your knowledge of the terms doctors use.

721 p.68 The word 'gastric' relates to which organ of the body?

722 p.130 What is medical jargon for thoroughly washing the hands before performing an operation?

723 p.130 What is the common name for sutures?

724 p.76 By having two eyes, we can judge distance and see things in three dimensions. What is this type of vision called?

725 p.52 Medical terms containing 'cardium' or 'cardio' refer to which organ?

726 p.46 The terms dermis, epidermis and subcutaneous relate to what part of the body?

727 p.62 What is the name of the scientific study of food and diet?

728 p.58 Pertussis is the medical name for a childhood disease that causes a severe cough. What is the disease called?

729 p.100 What does the 'neuro' in neurology mean?

730 p.62 What does the 'F' in the F-Plan Diet stand for?

Confused conditions

Unravel these anagrams of diseases and disorders.

731 p.68 IRRIGATE STETSON Medical term for stomach upset.

732 p.68 CAMELOT CRUSH Results in a severe bellyache (two words).

733 p.70 DONKEY INSET Hard cause of renal colic (two words).

734 p.74 MAZE RELISH Dementia of old age.

735 p.118 NAPKIN ROSS Causes shakes and walking problems.

736 p.100 SMALL SENTINEL Affects thought and brain function (two words).

737 p.114 LAVENDER FRUGAL Sometimes referred to as the 'kissing disease' (two words).

738 p.112 SESAME YIELD Tick-borne infection, especially in USA (two words).

739 p.112 SOFA CURL 'King's evil', a form of tuberculosis.

740 p.116 SACRED NOISE Tumours where cancer has spread.

Tuned in

The answers all relate to popular songs.

741 p.106 According to Peggy Lee, what symptom do you give her?

742 p.112 Love is the source of what, according to Van Halen on the 1988 album *OU812*?

743 p.92 In their hit song, what were Simply Red holding back?

744 p.120 Which Liverpool pop group, which included poet Roger McGough, had a hit in 1968 with 'Lily the Pink', whose medicinal compound was 'most efficacious in every case'?

745 p.66 'Food, Glorious Food' was a song from a musical based on which novel by Charles Dickens?

746 p.84 'Birds do it, bees do it, even educated fleas do it; Let's do it, let's…' what?

747 p.94 Which band was 'Dancing in the Street' in 1964?

748 p.96 What was Bing Crosby first dreaming of back in 1942?

749 p.134 Which British rock group recorded an album (and song) called 'My Iron Lung'?

750 p.100 Which 1980s ska-pop group had hits including 'Baggy Trousers' and 'Driving in My Car'?

Physical forces

Things that oppose, complement, neutralise...

751 p.40 What does the body contain more of: sodium or potassium?

752 p.56 What is the name of the blood proteins produced by some white blood cells to neutralise bacteria?

753 p.56 What is the term for a protein on the surface of a cell or organism that marks it as 'foreign'?

754 p.74 'Sympathetic' and 'parasympathetic' nerves act in opposite ways in what part of the nervous system?

755 p.80 When you scratch a mosquito bite to relieve the itching, what is the medical term for what you are doing?

756 p.94 Medically speaking, what is the opposite of aerobic?

757 p.98 In the psychological learning process known as 'reinforcement', what two techniques are used to stimulate learning?

758 p.122 In Chinese medicine, two opposing but complementary forces are at work. What are they called?

759 p.136 In drug biochemistry, what is the opposite of an antagonist?

760 p.138 To what group of drugs do aciclovir and zidovudine (AZT) belong?

**For answers and more facts go to the page given below each question number.
For quick answers to complete quizzes 72 to 76 go to page 149.**

QUIZ 76 IN YOUR STRIDE

Close-up body parts

Match these images with the list of body parts below.

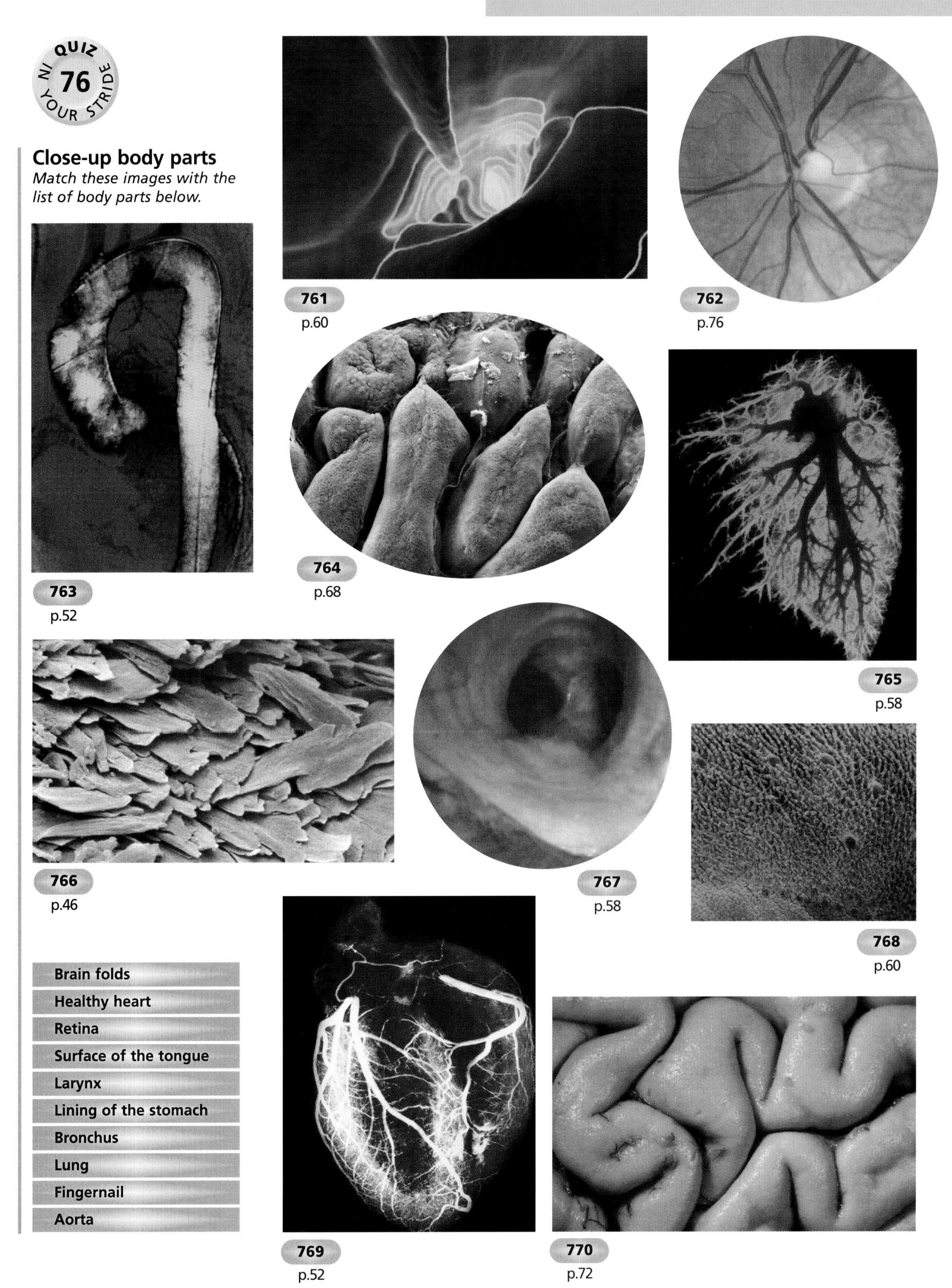

761 p.60

762 p.76

763 p.52

764 p.68

765 p.58

766 p.46

767 p.58

768 p.60

769 p.52

770 p.72

- Brain folds
- Healthy heart
- Retina
- Surface of the tongue
- Larynx
- Lining of the stomach
- Bronchus
- Lung
- Fingernail
- Aorta

True or false

Decide whether these statements are correct.

771 p.46 Fingernails grow up to four times as fast as toenails.

772 p.90 Very young girl babies tend to cry more than boy babies.

773 p.60 Teeth are outgrowths of the jawbones, and are made of bone.

774 p.118 Gout is always caused by drinking too much alcohol.

775 p.46 All types of skin cancer are mostly caused by excessive sun exposure.

776 p.88 In newborn babies, jaundice is common and is not usually a sign of anything serious.

777 p.72 The left-hand side of the body connects with the right-hand side of the brain.

778 p.62 Fruit juice is a good source of roughage, or fibre.

779 p.114 Antibiotics cure influenza.

780 p.46 Ringworm is caused by a minute worm living in the skin.

Know your body

A quiz on the workings of the human body.

781 p.106 'Normal' body temperature is 98.6°F. What is this in centigrade or Celsius?

782 p.54 What is the name of the red pigment that gives blood its colour?

783 p.80 The hard palate separates which two cavities in your head?

784 p.58 Breathing depends on the diaphragm and which other set of muscles?

785 p.110 What body fluid would a doctor sample to test your cholesterol levels?

786 p.84 A small gland at the base of the bladder may cause men difficulty in urinating if it becomes enlarged. What is it?

787 p.58 What gas is 100 times more plentiful in the air you breathe out than the air you breathe in?

788 p.52 The body's longest veins are the great saphenous veins. Where are they?

789 p.76 The optic nerves lead to the brain from where?

790 p.88 Name the two main components of the afterbirth.

Who, what or where?

A series of questions with a historical theme.

791 p.114 Spain (1918), Asia (1957), Hong Kong (1968) and Russia (1977) were the starting points for what?

792 p.126 What did Joseph Bazalgette build in London in the 19th century that had a huge positive impact on the health of the population?

793 p.112 Sir Ronald Ross received a Nobel prize in 1902 for discovering how mosquitoes transmit what killer disease?

794 p.40 Which Italian artist sculpted the larger-than-life figure of *David*?

795 p.120 Probably the most famous book of herbal remedies, first published in 1653, is still in print. Who wrote it?

796 p.78 Which great composer lived about 200 years too early to benefit from a cochlear implant?

797 p.102 What 1895 invention means that a doctor can literally see through you?

798 p.126 Dr John Snow is remembered for linking a water pump in Soho with which disease?

799 p.98 Frenchman Alfred Binet was a pioneer of which form of testing?

800 p.132 What kind of implant operations were perfected around 1960 by British doctor John Charnley?

Word games

Unravel the anagrams for a variety of medical terms.

801 p.40 MOONISH APES Mankind, scientifically speaking (two words).

802 p.42 OILY GHOST The study of tissues.

803 p.42 COSTLY MAP Jellylike content of cells.

804 p.52 TARA HACKETT Major cause of death (two words).

805 p.62 HUGO RAGE Another name for dietary fibre.

806 p.62 STEVE BAGEL Good source of previous question.

807 p.62 CEASE, LIAISE DOC Dietary intolerance disorder (two words).

808 p.122 PURE SUCRASE Ancient needle-free Chinese therapy.

809 p.126 SPURIOUS TALE Pioneer of rabies vaccination (two words).

810 p.136 CHARM APOLOGY Study of drugs.

For answers and more facts go to the page given below each question number.
For quick answers to complete quizzes 77 to 83 go to pages 149 and 150.

Who's who?

A round of questions with a people theme.

811 p.46 Who dreamt of living 'in a nation where [people] will not be judged by the color of their skin'?

812 p.54 Which World War II general was known as 'Old Blood and Guts'?

813 p.58 What is the term – named after a British physiologist – of the cycle of chemical reactions that releases energy from food molecules?

814 p.80 Merkel's discs, Meissner's and Pacinian corpuscles, and Krause's end-bulbs sense what?

815 p.82 For what disease did Frederick Banting, John Macleod and others find the chemical remedy in 1922?

816 p.94 Who, in July 1998, became the fastest woman on two legs?

817 p.98 Which American behavioural psychologist promoted the use of teaching machines for 'programmed learning'?

818 p.124 Who developed a system called eurythmy?

819 p.124 William Bates developed a series of self-help exercises to improve what?

820 p.144 Who was the woman who contributed greatly to discovering DNA's structure, but died before Crick, Watson and Wilkins won the 1962 Nobel prize?

Straight A's

The following answers all begin with the letter A.

821 p.44 What is the name for the genetic disorder which results in very short, misshapen legs and arms?

822 p.46 What is the name of the sweat glands in the armpits?

823 p.56 Rheumatoid arthritis and lupus erythamatosus are examples of what type of disease, in which the body attacks itself?

824 p.62 What are the chemical components of protein molecules?

825 p.82 Addison's disease and Cushing's syndrome are manifestations of problems with which glands?

826 p.40 Which goddess was the first to be represented as a full-size nude in Western art?

827 p.116 In the heart disease atherosclerosis, fatty deposits build up in the walls of the coronary arteries. What are these deposits called?

828 p.124 What is the term for the self-therapy that involves constantly repeating 'positive' phrases to oneself?

829 p.124 What philosophical system was developed by Rudolf Steiner, and later used as a basis for a holistic system of therapy?

830 p.128 What is the term for surgery in inherently germ-free conditions?

Who's counting?

Select the correct option from the four possible answers.

831 p.66 How many kilocalories does it take to put on 450 g (1 lb)?
A 3000 kcal B 4000 kcal C 5000 kcal D 6000 kcal

832 p.54 Blood plasma consists of roughly what proportion of water?
A 60 per cent B 70 per cent C 80 per cent D 90 per cent

833 p.100 Mental illness of some kind affects about how many people at some time during their lifetime?
A 1 in 3 B 1 in 5 C 1 in 10 D 1 in 20

834 p.50 In women muscles account for what percentage, on average, of body weight?
A 10 per cent B 20 per cent C 35 per cent D 45 per cent

835 p.46 On average the hairs on your head stop growing after?
A 2 years B 5 years C 10 years D 15 years

836 p.42 The shortest-lived cells in the human body die after about?
A Three days B Three weeks C Three months D Three years

837 p.72 The brain uses what proportion of the body's oxygen supply?
A 2 per cent B 5 per cent C 10 per cent D 20 per cent

838 p.88 On average, how many boys are born for every 100 girls?
A 96 B 99 C 106 D 113

839 p.92 People with Werner's syndrome suffer from what?
A A heart defect B Hypochondria C Schizophrenia D Premature ageing

840 p.92 Alzheimer's and other forms of dementia affect about what proportion of people over the age of 80?
A 1 in 20 B 1 in 10 C 1 in 5 D 1 in 3

In hospital

The answers in this round all have a hospital connection.

841 What is the name of a room in a hospital where surgical operations are performed? p.130

842 In a hospital, what is an autoclave used for? p.130

843 What is the general term for rehabilitation therapy after an injury, operation or illness, aiming to restore mobility and strength? p.124

844 What is the common name for one of the oldest and commonest tissue transplants, which doesn't involve opening up the body? p.128

845 In which department of a hospital would you find patients hooked up to machines constantly monitoring their heart rate, blood pressure and blood oxygen? p.134

846 What type of scanning is often used to observe an unborn baby? p.108

847 What is the name for a small, razor-sharp surgical knife used to cut tissues? p.130

848 In what general type of surgery is a heart-lung machine used? p.134

849 What is the common term for the process of childbirth? p.88

850 What was the nickname for the Hospital of St Mary of Bethlehem in London – one of the first asylums for the mentally ill – which became a term for chaos? p.100

True or false?

Decide whether the following statements are correct.

851 The left-hand half of the brain is usually dominant in controlling speech and writing. p.72

852 It is not possible for someone to have more than two sets of natural teeth in their lifetime. p.60

853 Down's syndrome is hereditary. p.44

854 A hot bath or sauna is as good as warm-up exercises before an aerobics session. p.94

855 Men, on average, have more red blood cells per litre of blood than women. p.54

856 By hanging loose from the body, the scrotum keeps the testicles cool, facilitating sperm production. p.84

857 In curly hair, the individual hairs have a round cross-section: the rounder the hair, the curlier it will be. p.46

858 Diet is not a factor in causing strokes. p.118

859 Identical twins have identical fingerprints. p.46

860 Chickenpox and shingles are caused by the same virus. p.114

Diagnosis

Name the diseases caused by the following organisms.

861 *Candida albicans* p.112

862 *Clostridium botulinum* p.112

863 *Clostridium tetani* p.112

864 *Corynebacterium diphtheriae* p.112

865 *Entamoeba histolytica* p.112

866 *Mycobacterium leprae* p.112

867 *Neisseria mengitidis* p.112

868 *Salmonella typhi* p.112

869 *Vibrio cholerae* p.112

870 *Yersinia pestis* p.112

Pot luck

Have a go at this mixed bag of teasers.

871 The names connective, epithelial, muscular and nervous all refer to what in the body? p.42

872 British doctor Edward Jenner deliberately exposed patients to a mild dose of which disease? p.56

873 What orange substance, found in carrots, is turned by the body into vitamin A and protects against night-blindness? p.64

874 Agar is a food additive used in ice cream and many other foods. What does it come from? p.66

875 What is the general name for a substance injected into the body so that soft parts show up as shadows on an X-ray? p.108

876 What is the term for a radioactive marker used to detect abnormal DNA due to a genetic defect? p.110

877 What is the medical term for diseases in which tissues or organs deteriorate and lose their function? p.118

878 What are the two most common types of cancer in children? p.116

879 In the Middle Ages, some diseases were blamed on excess blood in the diseased organ. What word was used for this excess? p.120

880 What type of surgical operation is a lithotomy? p.130

For answers and more facts go to the page given below each question number.
For quick answers to complete quizzes 84 to 88 go to page 150.

Science who's who

Can you identify these medical scientists by matching the names to the faces?

- Andreas Vesalius
- William Harvey
- Louis Pasteur
- Elizabeth Garrett Anderson
- Nettie Maria Stevens
- Sir Alexander Fleming
- James Watson
- Dr Gertrude Elion
- Christiane Nuesslein-Volhard
- Christiaan Barnard

882
p.44

881
p.102

885
p.44

883
p.106

884
p.102

887
p.102

886
p.144

890
p.102

888
p.44

889
p.74

Medical mix up
Use the clues to unravel the anagrams.

891 p.46 ORGANISING TOWLINE Painful condition for the chiropodist (two words).

892 p.54 SAMBA DOLLOP The liquid part of blood (two words).

893 p.64 MAN VISIT Foods you need for health.

894 p.72 NODAL CRISP Runs the length of the backbone (two words).

895 p.88 NOT NARCOTICS Painful parts of labour.

896 p.104 ECONOMY MERGER American term for hospital A&E (two words).

897 p.124 YAMAHA REPORT Massage treatment using fragrant oils.

898 p.126 VATICAN COIN Injection to prevent infection.

899 p.96 A LEMON TIN Hormone which regulates the body's 'clock'.

900 p.138 THIAMINE SAINT Treatment for allergies.

Prevention and cure
Ten questions on different kinds of treatment.

901 p.56 So-called 'memory' B-cells and T-cells are the basis of what common preventative medical procedure?

902 p.76 What instant remedy can you buy over the counter for presbyopia?

903 p.116 What is the most drastic treatment used for serious cardiomyopathy?

904 p.116 What everyday drug can help to thin the blood and prevent blood clots?

905 p.124 One controversial form of osteopathy involves massaging the skull. What is it called?

906 p.130 Give one of the two alternative terms for an injection of anaesthetic that numbs the lower part of the body, often from the waist down?

907 p.134 What radioactive isotope of a 'c' metal is often used to destroy tumours by implanting small 'seeds' in the body.

908 p.134 Apart from X-rays, name one other type of radiation, produced by a linear accelerator, that is used to kill cancer cells.

909 p.138 In relation to chemotherapy, what does 'cytotoxic' mean?

910 p.140 What is different about a transgenic transplant?

What's wrong?
A round about medical conditions and ailments.

911 p.62 What food substance, found in wheat and rye, should people with coeliac disease avoid?

912 p.74 Encephalitis is the inflammation of what?

913 p.78 What is the medical term for ringing in the ears?

914 p.78 What condition may a cochlear implant be used to treat?

915 p.86 What is the name of the condition in which an embryo starts to develop in a Fallopian tube instead of in the uterus?

916 p.84 Most cases of cervical cancer are caused by a virus, which also causes what skin condition?

917 p.92 Which one of the following does not directly influence how long you are likely to live: exercise; genetics; height; blood pressure; smoking; weight?

918 p.100 A German baron was so famed for his exaggerated stories that a syndrome in which a patient feigns symptoms in order to get medical treatment is named after him. What is his name?

919 p.100 By what name is the mental illness bipolar disorder better known?

920 p.116 What is the name given to cancer of the tissues that line organs and form skin and membrane.

That *is* the question!
The answers in this round all end in 'is'.

921 p.80 What is the name for an inflammation of the mucous membrane that lines the nose?

922 p.48 What is the proper name for 'housemaid's knee'?

923 p.48 What is the proper name of the disease that causes 'dowager's hump'?

924 p.50 What is the proper name for tennis elbow and similar conditions?

925 p.70 What disease, often a result of excess alcohol consumption, causes scarring and damage to the liver?

926 p.70 What is the name of the blood-cleansing process for which a kidney machine is used?

927 p.60 What flap of cartilage prevents you choking every time you swallow food or drink?

928 p.58 What is the name for infection and/or inflammation of the bronchi of the lungs?

929 p.118 Rheumatoid arthritis and gout are two major forms of arthritis. What is the third?

930 p.74 Infection of the membranes surrounding the brain is called what?

For answers and more facts go to the page given below each question number.
For quick answers to complete quizzes 89 to 95 go to page 150.

True or false?

Decide whether these statements are correct.

931 p.132 The first successful transplant of the cornea of the eye took place nearly 50 years before the first successful kidney transplant.

932 p.72 Nerve signals travel at an average speed of 50 m/s (165 ft/sec).

933 p.140 A form of genetically engineered maize can produce a contraceptive that could be eaten as cornflakes.

934 p.48 All joints allow movement of bones.

935 p.68 Without a protective coating of mucus, the stomach and intestines would digest themselves.

936 p.86 In fertilisation, several sperm enter the egg, but only one of them survives.

937 p.40 Women have a lower centre of gravity than men.

938 p.136 Many modern drugs are designed by computer.

939 p.142 Infant deaths from vehicle accidents in Britain were more than twice as high in 1995 as in 1949.

940 p.116 A typical tumour has more than a billion cancer cells before it is detected.

Movie time

Ten questions about films with a medical connection.

941 p.116 A 2000 movie starring Julia Roberts as a real-life campaigner against corporate pollution.

942 p.68 Which film starred Meryl Streep and Jack Nicholson as two journalists?

943 p.84 *… lies and videotape.* A 1989 film starring Andie MacDowell.

944 p.100 A 1999 film starring Winona Rider depicting life in a mental institution.

945 p.96 A 1946 Howard Hawks movie starring Humphrey Bogart as private eye Philip Marlowe alongside Lauren Bacall.

946 p.58 The original French film, made in 1960 by Jean-Luc Godard and starring Jean-Paul Belmondo, was *A bout de Souffle*. What was the title of the 1983 US remake?

947 p.76 A 1978 movie in which Faye Dunaway played a fashion photographer who foresees a series of murders.

948 p.94 The British-made 1981 film about athletic achievements at the 1924 Paris Olympics.

949 p.96 A 2002 film set in Alaska and starring Al Pacino as an LA cop.

950 p.86 Which 1999 film, starring Hugh Laurie and Joely Richardson, is about a couple trying to conceive?

Body parts

Select the correct option from the four possible answers.

951 p.74 Which nerve cells carry signals to the muscles?
A Active neurons | B Conscious neurons
C Motor neurons | D Sympathetic neurons

952 p.74 Which part of the body is affected by Bell's palsy?
A An arm | B The buttocks
C The face | D A leg

953 p.72 A nerve fibre 'fires' at about what voltage?
A 0.01 volts | B 0.1 volts
C 1 volt | D 10 volts

954 p.76 What is the proper name for the 'blind spot' in each eye?
A Asensory disc | B Neural disc
C Occluded disc | D Optic disc

955 p.82 Between them, calcitonin and parathyroid hormone regulate the body's content of which mineral?
A Calcium | B Iron
C Potassium | D Sodium

956 p.96 What is the maximum time that anyone has been observed to stay awake?
A 7 days | B 9 days
C 11 days | D 13 days

957 p.98 Which word describes mental processes such as reasoning?
A Affective | B Cognitive
C Conscious | D Intelligent

958 p.98 What term is used for the most severe mental illnesses?
A Ideopathic | B Neurosis
C Psychopathic | D Psychosis

959 p.128 Which was first to be introduced as an anaesthetic?
A Chloroform | B Ether
C Halothane | D Nitrous oxide

960 p.128 What discovery in 1900 indirectly made major surgical operations possible?
A Anaesthesia | B Blood groups
C Cauterisation | D Shock

True or false?

Decide whether the following statements are correct.

961 p.96 Except during dreams, brain activity stops when you are asleep.

962 p.90 Boys tend to have better spacial awareness than girls.

963 p.60 Build-up of dental plaque is the major cause of gum disease.

964 p.94 Yoga is a very good way of improving stamina.

965 p.88 The Guthrie test involves pricking the heel of a new-born baby.

966 p.40 The shortest known adult human was less than 0.6 m (2 ft) tall.

967 p.52 Smoking, high blood pressure and athero-sclerosis are all major causes of heart disease.

968 p.118 There are more than 100 different forms of arthritis.

969 p.62 White bread contains no roughage, or fibre.

970 p.50 In body-building, diet is as important as exercise.

Pot luck

A general round of brainteasers.

971 p.86 What is the medical term for the womb?

972 p.90 What is the name for the 'soft spots' in a baby's skull?

973 p.120 Which Christian church is based on a belief in faith healing?

974 p.124 What is the name of the therapy that uses electronic instruments to measure unconscious bodily changes in order to help the patient learn to control them?

975 p.124 What alternative technique harnesses a person's imagination?

976 p.126 What was the popular name for the putrid smell of the River Thames in the hot summer of 1858?

977 p.132 What is the difference between an implant and a transplant?

978 p.132 What term is commonly used for electronic parts that mimic natural body functions?

979 p.134 Give either of the two terms used for a machine that aids breathing during and after surgery.

980 p.144 The first Nobel prize for medicine, awarded in 1901, was for a serum to treat which disease, often fatal to children before a vaccine was developed?

Name the singers

Ten musical questions with a slim link to body and health.

981 p.94 Which singer was 'Born to Run'?

982 p.64 Who had a hit with 'Banana Republic' in 1980?

983 p.74 Which band had a hit with 'The Reflex'?

984 p.74 Who sang 'Memories, pressed between the pages of my mind'?

985 p.64 'Rip it Up' was a hit for which British band?

986 p.78 Who sang, in 1980, 'Did I hear you say you love me?'

987 p.58 Which band will be watching you, 'Every breath you take'?

988 p.54 Who sang 'Mud, mud glorious mud,/Nothing quite like it for cooling the blood'?

989 p.88 In 1938 who had a No. 1 hit singing 'You must have been a beautiful baby… 'Cause baby look at you now' – later picked up by Bobby Darin and by the Dave Clark Five?

990 p.76 Who was 'skipping and a-jumping/In the misty morning fog' with his 'brown-eyed girl'?

Old times

Name the modern equivalent of these old medical names.

991 p.126 Enteric fever

992 p.118 Apoplexy

993 p.114 Breakbone fever

994 p.58 Consumption

995 p.116 Dropsy

996 p.74 Falling sickness

997 p.114 Grippe

998 p.114 Infantile paralysis

999 p.118 Screws

1000 p.58 Winter fever

**For answers and more facts go to the page given below each question number.
For quick answers to complete quizzes 96 to 99 go to page 150.**

BODY AND HEALTH

Essential facts, figures and other information on the human body – how it works, what is good for it, what can go wrong, and the medical treatments available to remedy ill health.

QUESTION NUMBER

The numbers or star following the answers refer to information boxes on the right.

ANSWERS

92	**50-60 per cent** ❸
134	**D: 635 kg (1400 lb)** – Jon Brower Minnoch (1941-83)
275	**The nails** (the body has less than 5 g of iron) ❸
281	**True** (*Homo neanderthalensis*) ❶ ❷
412	**Feet** – in 'Your Feet's Too Big' (1939)
527	**D: A thin, gangly person** ❺
552	**Body mass index** ❼
712	**A: Invertebrates** (we have backbones) ❶
751	**Potassium** ❸
★ 794	**Michelangelo**
801	***Homo sapiens*** ❶ ❷
826	**Aphrodite** ★
937	**True** ❹
966	**True** – Gul Mohammed (1957-97) of New Delhi is 57 cm (22½ in) tall

The amazing body machine

Core facts ❶

◆ Human beings are **mammals** – warm-blooded animals with a backbone and hair, rather than scales or feathers, whose females suckle their young. Like apes and monkeys, humans belong to the **primate** order.
◆ Humans are members of the biological family ***Hominidae*** (hominids), with a scientific name *Homo sapiens*, meaning 'wise human'.
◆ The human body is made up of **chemical elements**, and is the most complex structure in the living world: its sophisticated brain sets it apart from all other creatures.
◆ Genetics, sex differences and lifestyle choices create a huge variety of human **body shapes**.

Human evolution ❷

Humans developed from ape-like ancestors over a period of 5-10 million years. Their evolutionary changes led to:
◆ a larger, more complex **brain**;
◆ a finer jaw with smaller **teeth**, through adapting from a vegetarian to a mixed diet;
◆ a **voice box** sitting lower in the throat, making complex speech possible;
◆ an **upright stance**, with long leg bones, locking knee joints and a small pelvis;
◆ increased flexibility and strength in the **hands**, enabling the skilled use of tools.

Two genera (species groups) of hominids have been identified: *Australopithecus* (extinct by 1 million years ago) and *Homo*, of which humans are the sole survivors. Modern humans, ***Homo sapiens***, evolved in Africa 200 000-140 000 years ago and reached Europe about 100 000 years ago.

Body chemistry ❸

The body – like its fuel, food – is mainly composed of water, fat, protein and carbohydrate. The remainder is made up of chemicals such as minerals, salts and acids.

Chemical component	Approximate percentage of body weight
Water	50-60 per cent; more in men than women.
Fat	10-50 per cent, including 1 per cent of 'essential' fats, needed to maintain cells. More than 25 per cent in men and 30 per cent in women is unhealthy
Protein	18 per cent
Carbohydrate	0.5 per cent
Major minerals	**Average percentages**
Calcium	1.47 per cent
Phosphorus	1 per cent
Potassium	0.34 per cent
Sulphur	0.24 per cent
Sodium	0.15 per cent

★ 794

The body in art ★

Artistic representations of humans are found in paintings and figurines from the Palaeolithic period, 40 000 to 10 000 years ago. By 1500 BC, Egyptian and Mycenean artists were creating images of ideal beauty. Greek sculptors continued the tradition, notably **Praxiteles**, whose *Aphrodite of Cnidus* (c.350 BC) was the first known full-size female nude. Greek influence can be seen in **Michelangelo**'s *David* (1501-4) and **Canova**'s *Three Graces* (1813-16), but 20th-century artists turned to more abstract forms as in **Picasso**'s *Three Dancers* (1925).

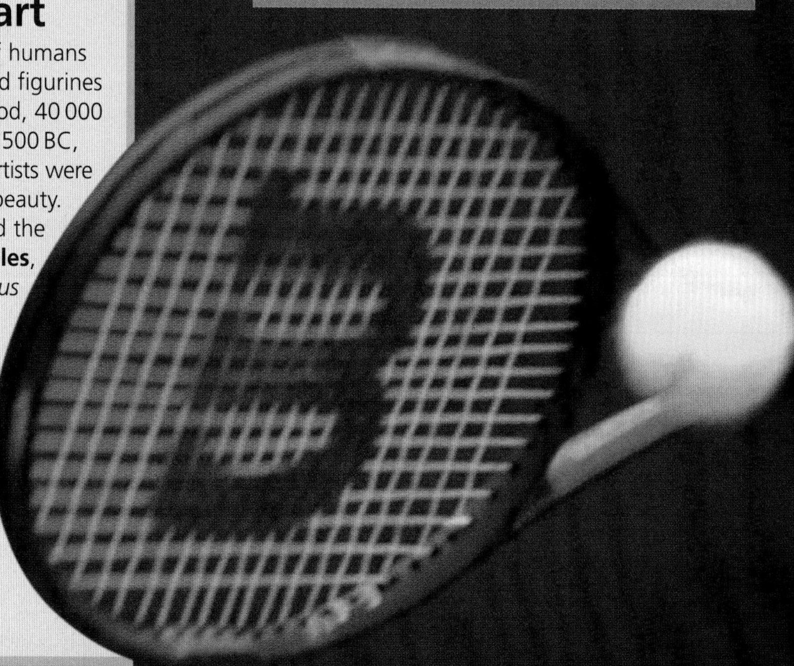

Male and female

❹

Women have, on average, more body fat than men, less muscle mass, and lighter, smaller bones – in particular, shorter and less muscular legs. They have lower individual and total muscle strength, a smaller heart and lungs in relation to body size, and a lower metabolic rate.

In top athletic competition, men's muscle mass, strength and aerobic capacity generally allows them to out-perform women. Women have a lower centre of gravity, around the hips, which confers an advantage in throwing events, but makes jumping events more difficult.

FINE PHYSIQUE Venus Williams, winner of Wimbledon in 2000 and 2001, has an advantage that gives her slightly more range than other women tennis players – she is 1.85 m (6 ft 1 in) tall.

Extreme body shapes

❺

Most bodies are a mixture of three main body types (somatotypes):

- **Ectomorph**, narrow and angular with spindly limbs and little muscle or fat;
- **Endomorph**, with heavy build, rounded head and abdomen and thick body fat;
- **Mesomorph**, narrow-hipped, muscular, with broad shoulders and a large head.

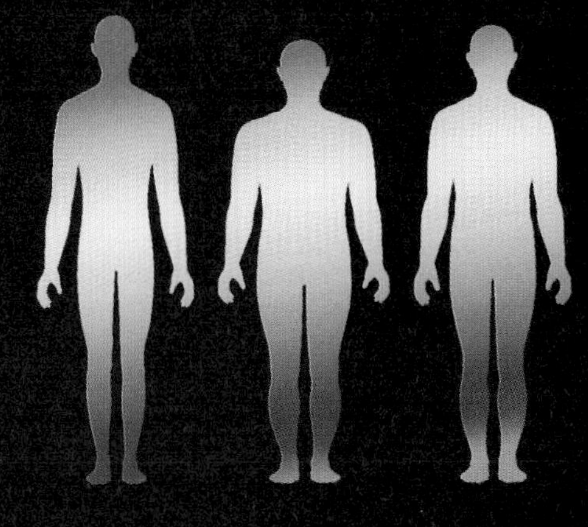

Tie-breaker

❻

Q In which country is the world's shortest ethnic group found?

A Democratic Republic of Congo (formerly Zaire). The Mbutsi people average 1.36 m (4 ft 5 in) tall. The tallest people are the Tutsi of Rwanda and Burundi – adult men average 1.83 m (6 ft) tall.

A healthy weight

❼

Body mass index (BMI) is a measure of the ideal body-weight range for a particular height, and is calculated using the formula:

$$\text{BMI} = \frac{\textit{Weight in kilograms}}{(\textit{Height in metres})^2}$$

For example, the calculation for a person weighing 80 kg and measuring 1.7 m tall is 80 ÷ 2.89 (1.7 x 1.7), giving a BMI of 27.68.

The ideal BMI falls between 20 and 25. Over 25 indicates that a person is overweight, and over 30 is obese; below 20 is underweight. These factors influence health, as can the distribution of body fat. A waist measurement of more than 102 cm (40 in) for men or 89 cm (35 in) for women indicates an increased risk of heart disease.

QUESTION NUMBER

The numbers or star following the answers refer to information boxes on the right.

ANSWERS

- **151** Nerve (the others are organs) ❶
- **391** Inside a cell ❷
- **392** Cells ❶ ❷ ❼
- **401** C: 10 trillion ❶
- ★ **451** Ovum cell ❹
- **452** Skin cell ❺
- **453** Red blood cell ❹ ❺
- **454** Nerve cell ❹ ❺ ★
- **455** Fat cell ❹ ❺ ★
- **456** Liver cell ❹
- **457** Cardiac muscle cell ❺
- **458** Epithelial cells ❺
- **459** White blood cell ❹ ❻ ★
- **460** Sperm cell ★
- **521** A: Monks' cells ❸
- **802** Histology ❼
- **803** Cytoplasm ❼
- **836** A: Three days ❹
- **871** Tissues ❺

Cells and tissues

Core facts ❶

◆ The human body is made up of **cells** – the smallest units capable of carrying on the processes of life. An average adult is made up of more than 100 000 billion cells, and yet began as a single cell: a fertilised egg.

◆ Cells vary widely in size, shape and function, but all have a similar basic **structure** and similar internal units, called **organelles**, which carry out metabolic functions.

◆ Groups of similar cells form **tissues**, which work together to carry out particular functions. Examples include muscle and nerve tissue. Tissues combine to form organs such as the brain, heart, lungs and kidneys.

Cell structure ❷

The cells of complex organisms differ in their size, shape and function, but share similar internal characteristics. The cell membrane encloses a jelly-like fluid called **cytoplasm**, 75 to 80 per cent of which is water. Suspended within the cytoplasm are **organelles** – tiny structures which, among other things, manufacture and store **energy** from food molecules, synthesise **proteins** and **fats** and destroy **toxins**. Vital substances such as **antibodies**, **hormones** and **nutrients** pass between cells via receptor sites on the membrane.

The **nucleus** controls these processes; it also contains deoxyribonucleic acid (DNA), a substance which codes the **hereditary characteristics** of an organism.

Cell membrane A 'skin' of protein and fat layers regulates the materials passing in and out of the cell.

Endoplasmic reticulum Manufactures proteins and fats and transports materials around the cell through a system of tubes. Covered in tiny protein-synthesising organelles called ribosomes.

Golgi apparatus Stores and transports proteins and fats.

Nucleus Control centre of the cell, regulating chemical processes; contains most of the cell's hereditary material (DNA).

Mitochondria Powerhouses which burn food molecules to release energy.

Lysosomes Contain enzymes which digest proteins and fats, and break down toxins and debris.

WEIRD AND WONDERFUL ❸

The word 'cell' was coined by British scientist **Robert Hooke** (1635-1703). He thought that the rows of cork cells he saw through his primitive microscope looked like the tiny, bare monastery cells lived in by monks.

★ 451 ★

Cell shapes

Human cells vary from round, globular **fat cells** and **ova** (eggs) to extremely elongated, branching **nerve cells** (neurons) and tiny tadpole-like **sperm**. Many cells are box or column-shaped; others are flattened. Some cells can change their shape – **white blood cells**, for example, can squeeze through the walls of capillaries (the finest blood vessels) to reach a site of infection, or fold up to engulf disease organisms such as bacteria.

BACKGROUND IMAGE Cells in the retina of the eye contain granules of pigment.

Cell record-breakers

Most cells are too small to see, averaging about 0.01 to 0.02 mm (one-hundredth to one-fiftieth of a millimetre) across. But the human body contains about 200 types of cell, some of which are very different from one another.

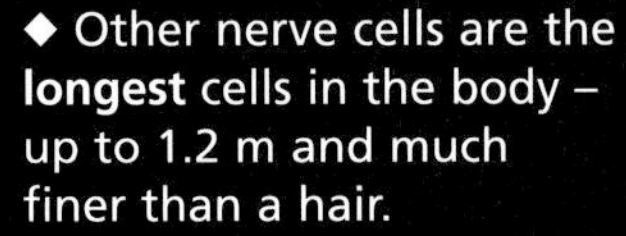

- The **biggest** human cells are female egg cells, or ova. An ovum is just visible to the naked eye and measures about 0.035 mm in diameter.
- Some of the **smallest** cells in the human body are nerve cells found in the brain, which measure only 0.005 mm across.
- Other nerve cells are the **longest** cells in the body – up to 1.2 m and much finer than a hair.
- The **shortest-lived** cells are those lining the inside of the mouth, which die and are replaced after only three days. White blood cells live about two weeks, red blood cells about four months, and liver cells about 18 months.
- Nerve cells are the **longest-lived**. They form in the foetus in the womb and are not replaced if they die, so some last a lifetime.

ADIPOSE TISSUE Fat cells store energy, insulate against the cold and protect the body from injury.

Tissue types

The human body is made up of four basic types of tissue:

- **Connective tissue** Joins or supports the body's tissues or organs, and consists of various cell types. Examples include bone, cartilage and adipose tissue (made up of fat cells). Blood is also considered to be connective tissue, with fluid between its cells.
- **Epithelium** Tightly packed sheets of cells forming membranes. Epithelium covers the body's internal and external surfaces, forming the skin and mucous membranes.
- **Muscle tissue** Fibres that contract when stimulated by nerve signals, transforming chemical energy into mechanical energy.
- **Nervous tissue** Cells of the brain and nervous system, which transmit electrochemical signals.

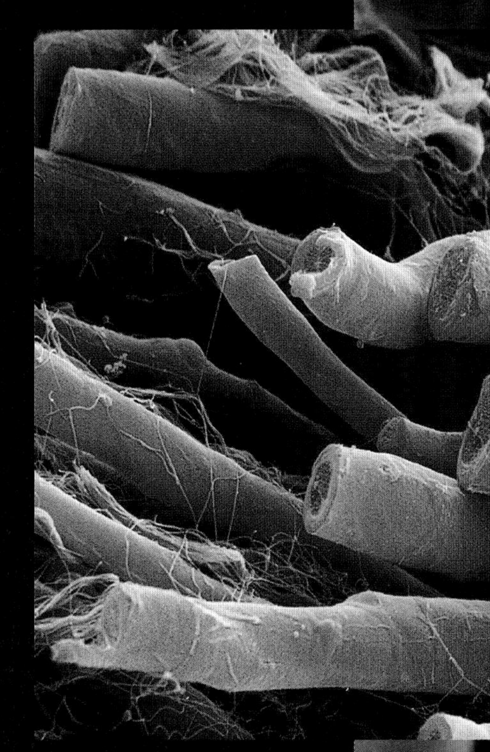

SENDING MESSAGES Axons (nerve fibres) in nerve cells carry electrical signals along one cell to the next.

Tissue typing

Tissue transplants activate the **human leucocyte antigen** (or HLA) system, by which white blood cells recognise protein molecules attached to the transplanted cells as 'foreign' and attack them. The genetic HLA profiles of donor and recipient must match closely to allow a successful transplant.

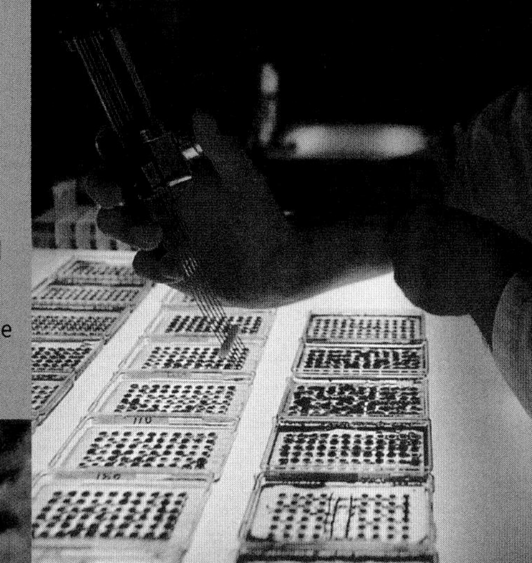

VITAL PROCESS Tissue typing matches donor and recipient tissues.

Some key terms

Cytology Study of the structure and function of cells.
Cytoplasm The jelly-like material inside the cell wall, excluding the nucleus, but including the organelles.
Eukaryotic cell A cell with hereditary material (DNA) contained within its nucleus.
Histology Study of tissues.
Organelle Any structure inside a cell with a specific metabolic function, such as the mitochondria.
Prokaryotic cell A cell with hereditary material (DNA) in its cytoplasm, rather than in the nucleus. Found in organisms such as bacteria and algae.
Protoplasm The entire internal contents of a cell – including the cytoplasm and the nucleus.

Tie-breaker

Q What is the name of the energy-containing substance synthesised in a cell's mitochondria?
A Adenosine triphosphate (ATP). The energy stored in ATP drives several metabolic processes, from the transport of substances in and out of cells (known as 'active transport') to muscular contractions.

Genes, chromosomes, DNA

QUESTION NUMBER

The numbers or star following the answers refer to information boxes on the right.

ANSWERS

Question	Answer
41	46 ❷
65	**B: Hyperlipidaemia** (high cholesterol) ❼
93	**Chromosomes** (*khroma* = colour, *soma* = body) ❶ ❷
148	**Blue eyes** (Sinatra was 'Ol' blue-eyes') ❹
181	**F: Grandson/G'father** – Russell Crowe/Stan Wemyss
182	**D: Mother/Dau** – Vanessa Redgrave/Joely Richardson
183	**E: Mother/Son** – Maggie Smith/Toby Stephens
184	**B: Son/Father** – Michael/ Kirk Douglas
185	**A: Brother/Brother** – Casey/Ben Affleck
186	**C: Brother/Brother** – Kieran/Macaulay Culkin
187	**I: Brother/Sister** – Warren Beatty/Shirley Maclaine
188	**G: Sister/Sister** – Kate/Minnie Driver
189	**J: Sister/Sister** – Louise Adams/Victoria Beckham
190	**H: Father/Son** – Donald/ Kiefer Sutherland
241	**Nucleus, mitochondria** ❶ ★
★ 256	**D: A human ancestor**
647	**C: 1953** ❻
650	**D: 2003** ❻
821	**Achondroplasia** ❼
853	**False** (caused by chromosome defect) ❸
882	**Nettie Maria Stevens** – heredity (1861-1912)
885	**Christiane Nuesslein-Volhard** – genetics (1942-)
888	**James Watson** (DNA; 1928-) ❻

Core facts ❶

◆ Children inherit visible characteristics from their parents such as skin and eye colour, and invisible characteristics such as genetic disease.
◆ The inherited characteristics of each human are determined by 30-40 000 **genes**, or units of heredity, which consist of segments of a chemical called **deoxyribonucleic acid (DNA)**.
◆ Most DNA is found in the nuclei of cells. In normal **cell division**, the DNA duplicates and is concentrated in the **chromosomes**. Each new cell has a duplicate set of chromosomes.
◆ Except in eggs and sperm, chromosomes occur in pairs. During **fertilisation**, one of each pair comes from each parent.

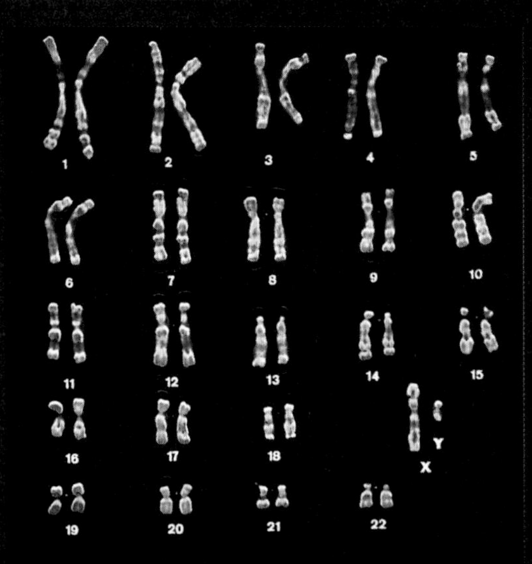

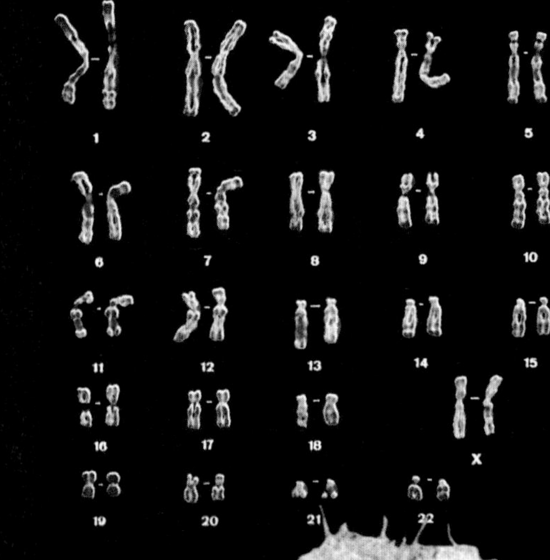

The dividing cell ❷

The microscopic photographs above show the human **karyotype** – the number and appearance of chromosomes in the nucleus of a male (left) and female (right) cell. It shows **46 chromosomes** in each, found mostly in pairs of the same size. Men have 22 matching pairs, plus a pair of odd chromosomes – a single **X** (from the mother's egg) and a smaller **Y** (from the father's sperm). Women have 23 pairs, including a pair of X chromosomes (one from each parent). The sperm determines the offspring's **sex** as it contains either an X or a Y chromosome.

SEX DETERMINATION Men inherit a Y chromosome (above) and an X chromosome (right) from their parents.

Chromosomal disorders ❸

Some chromosomal disorders are caused by having the wrong number of whole chromosomes, and arise because of defects in the sperm or egg before fertilisation.

Disorder	Features
Down's syndrome	Three no.21 chromosomes instead of the usual two.
Klinefelter's syndrome	Occurs in boys who have an extra X chromosome (XXY instead of XY). Causes failure to develop at puberty.
Turner's syndrome	Occurs in girls with one X chromosome (X instead of XX). Causes reduced growth.

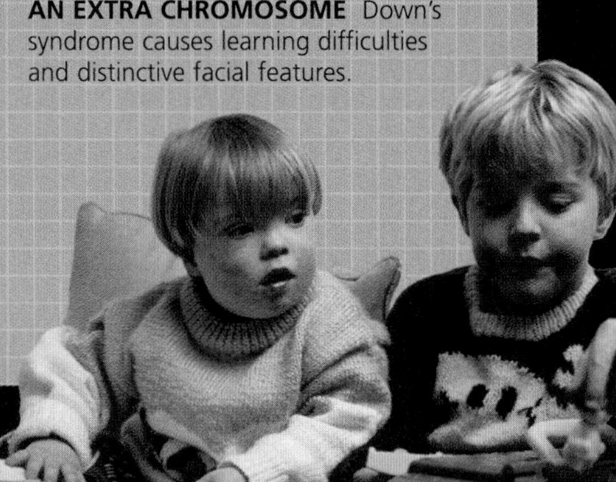

AN EXTRA CHROMOSOME Down's syndrome causes learning difficulties and distinctive facial features.

Dominant and recessive ❹

Each chromosome carries a set of **genes**, and each chromosome pair contains two versions of most genes – one from each parent. If both genes code for the same characteristic, such as blue eyes, the individual will possess that characteristic. But in many cases a gene pair codes for two versions of a characteristic, such as brown and blue eyes. In this case, the brown eyes gene is **dominant**, and the individual always has brown eyes, even if the gene is inherited from only one parent. The blue eyes gene is **recessive** – it only appears if both parents pass on a blue eyes gene.

FAMILY LIKENESS The inheritance of characteristics such as hair, eye and skin colour is governed by more than one gene.

GENETIC TESTING The sequence of parts of a person's DNA are almost unique, so an X-ray of DNA sequences from hair, skin or semen can be used for identification.

Mitochondria ★

Part of a cell's DNA can be found in its mitochondria, which are inherited from the mother. Mitochondrial DNA mutates at a rate of about 2 per cent every million years, and by measuring this rate in related creatures, scientists can estimate the date of a common ancestor. Today's humans trace back to **'Mitochondrial Eve'** – a female who lived 200 000 years ago.

WEIRD AND WONDERFUL ❺

Queen Victoria carried **haemophilia** in her genes. Her children married into royal families in Germany, Spain and Russia, passing the disorder on, and haemophilia became known as the **'royal disease'**.

TIMESCALE ❻

▶ **1865** Gregor Mendel publishes his work on plant heredity. He was ignored.
▶ **1869** DNA discovered in human cells.
▶ **1900** Mendel's work rediscovered.
▶ **1910** Thomas H. Morgan links genes with chromosomes.
▶ **1944-52** Oswald Avery and others prove that DNA determines heredity.
▶ **1951-3** Rosalind Franklin succeeds in taking X-ray photos of DNA.
▶ **1953** Francis Crick and James Watson discover the structure of DNA.
▶ **1960s** Sydney Brenner, Marshall Nirenburg and others discovered how genetic data is transferred.
▶ **1976** An artificial gene is inserted into a bacterium in the first successful trial of genetic engineering.
▶ **1977** Frederick Sanger deduces the sequence of units within DNA.
▶ **1982** First genetically engineered drug – insulin – approved for use.
▶ **1983** First genetic disease test developed.
▶ **2001** Human embryos first cloned.
▶ **2003** Human Genome Project publishes human DNA sequence.

Major genetic disorders ❼

If both parents have a disorder caused by a single **dominant** gene, there is still a 50:50 chance that their offspring will inherit two versions of the disease-free gene. **Recessive** disorders manifest themselves when an affected recessive gene is inherited from both parents. **Sex-linked** disorders are attached to one of the X or Y sex chromosomes.

Dominant disorders	Parts affected	Result
Achondroplasia	Bone growth	Short, misshapen limbs.
Huntington's chorea	Brain	Abnormal movements; dementia.
Hyperlipidaemia	Blood	High cholesterol; risk of blocked arteries.
Recessive disorders		
Cystic fibrosis	Lungs, digestion	Thick secretions; breathing problems.
Sickle-cell disease	Blood	Abnormal red blood cells block vessels.
Thalassaemia	Blood	Anaemia.
Sex-linked recessive disorders		
Haemophilia	Blood	Abnormal clotting; bleeding.
Red-green colour blindness	Eyes (retina)	Reduced ability to distinguish colours.

Skin, hair and nails

QUESTION NUMBER

The numbers or star following the answers refer to information boxes on the right.

ANSWERS

135 **B: 2 m² (21 sq ft)** ❶

136 **C: 4 mm** ❶

191 **False** (they have roughly the same number) ❻

277 **Footprints, iris pattern, or voice print** ❺

299 **Reddish** (accept red or orange) ❹

386 **False** (they have about the same number) ❹

402 **D: 2.5 million** ❼

414 **Lloyd Cole and the Commotions** – 1984

511 **Melanocytes** ❷ ❹

512 **Hair follicles** ❷

545 **Antipruritic** (drug to alleviate itching) ❽

699 **Paul Muni** – directed by Howard Hawks

726 **The skin** (from Greek *derma*, skin) ❷

★ 766 **Fingernail**

771 **True** ❼ ★

775 **True** ❽

780 **False** (caused by microsopic fungi) ❽

811 **Martin Luther King** – 'I have a dream' speech, 1963

822 **Apocrine glands** ❷

835 **C: 10 years** ❼

857 **False** (a round cross-section produces straight hair) ❸

859 **False** ❺

891 **Ingrowing toenails** ❶ ❼

Core facts ❶

◆ The skin is the body's **biggest organ**, with an area of about 2 m² (21 sq ft) in an adult male.
◆ **Skin thickness** varies between about 1.5 and 4 mm (1/16 and 1/6 in) – thinnest on the eyelids, thickest on the soles of the feet.
◆ The skin forms an almost waterproof cover for the body, **protecting** against chemicals, germs and harmful light rays and **preventing dehydration** by retaining moisture.
◆ The skin helps to regulate the body's **temperature**. It also manufactures **vitamin D** and contains nerves which transmit the sensations of touch, pain, heat and cold.
◆ **Hair** and **nails** are outgrowths of the skin.

Skin deep ❷

The outermost layer of the skin, the **epidermis**, ranges from 0.1 to 1 mm thick, and is covered with a layer of dead skin cells which constantly flake off. Pigment-producing melanocytes lie beneath the epidermis and determine the colour of the skin.

The next layer – the fibrous, elastic **dermis** – is four to ten times thicker than the epidermis, and contains the skin's most important structures. Blood vessels regulate temperature, cooling the body by dilating to allow more blood to flow or constricting to conserve heat. Hair follicles contain nerve endings sensitive to touch. Hair-erector muscles make hairs stand on end, trapping a warm layer of air next to the skin. Sebaceous glands lubricate the hair and skin with fatty secretions. Sweat glands cool the body, with ducts carrying sweat to the surface; apocrine sweat glands, mainly in the groin and armpits, produce sweat that smells when broken down by bacteria and acts as a sexual signal.

Beneath the dermis is the **hypodermis** – a layer mainly made up of fat cells which act as a food store and insulate the body.

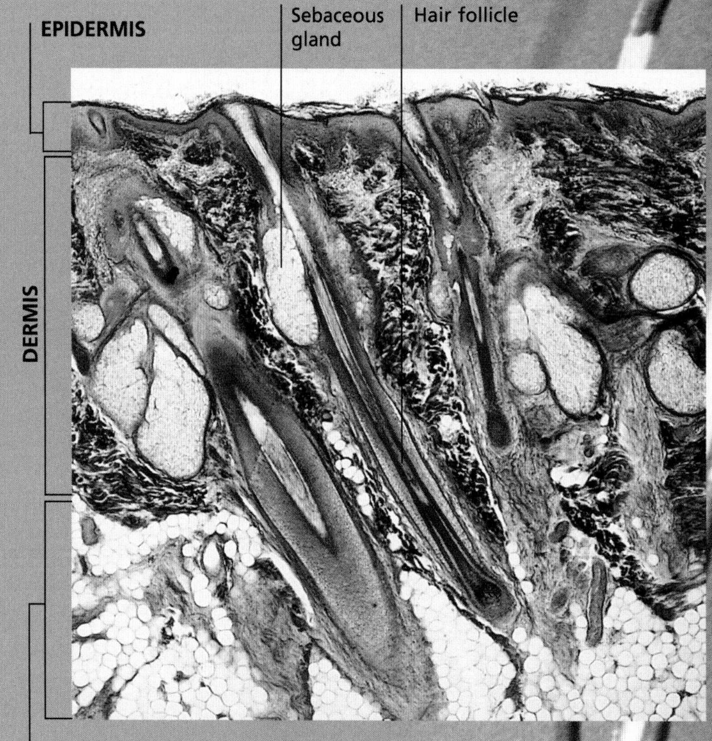

PROTECTIVE BARRIER The three layers of human skin can be seen in this cross-section.

Hair ❸

Hairs are made of dead cells, many of which contain a protein called **keratin.** The size and shape of the follicle determines the form of a hair. Round follicles produce **straight hair**; if some follicles are round and others are oval, the result is **wavy hair**. **Curly hair** is ribbon-like – caused by flattened, elongated follicles.

★ 766 ★

Fingernails

Nails grow from the epidermis and, like hairs, consist mostly of cells containing hard **keratin**. Apart from the growing area – the matrix or root at the base of the nail – nails are composed of dead cells. Their **pink colour** is caused by blood vessels in the skin underneath.

COLOUR CREATION Activity in pigment cells determines skin colour. Caucasian skin (left) has the least active cells; in Asian skin (middle) they are more active; in black skin (right) they are larger and active.

Skin pigmentation

❹

Skin and hair is coloured by different forms of a pigment called **melanin**: reddish phaeomelanin (mostly in hair) and dark brown eumelanin. They also contain a smaller amount of a yellowish pigment called **carotene**. Melanin is made by melanocytes – cells which exude coloured particles that are engulfed by skin and hair-producing cells. Dark and pale-skinned people have about the same number of melanocytes, but they are more active in dark skin.

Fingerprints

❺

In the womb, a permanent pattern of ridges develops on the skin of the lower surface of the fingers and toes, the palms and the soles of the feet. These patterns are unique to each individual – though the patterns of identical twins are about 95 per cent the same.

Fingerprint features

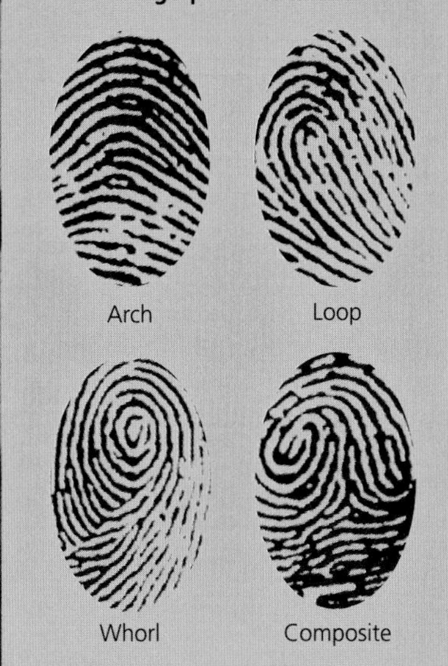

WEIRD AND WONDERFUL

❻

Humans – sometimes called 'naked apes' – have as many **body hairs** as other primates, but most hairs are shorter and finer. The development of intelligence enabled prehistoric naked apes to survive the ice ages.

LOW-MAINTENANCE HAIR Afro-Caribbean hair that is never combed naturally tangles into dreadlocks. Caucasian hair usually needs perming, backcombing or training to achieve the same effect.

Skin and hair stats

❼

◆ You have about 2.5 million sweat glands. The highest concentration is on the palms: 500 per cm^2 (over 3225 per sq in).
◆ In a year, you shed about 4 kg (9 lb) of dead skin and up to 30 000 hairs.
◆ In one week, fine hairs grow about 1.5 mm (1/16 in), coarse hairs 2.5 mm (1/10 in). Head hairs can grow for ten years or more, reaching up to 1 m (3 ft) in length, before the follicle becomes inactive and they fall out.
◆ Fingernails grow up to 5 cm (2 in) a year – up to four times as fast as toenails.

Skin disorders

❽

◆ **Skin infections** are caused by bacteria (boils and impetigo, for example), fungi (athlete's foot and ringworm – not a worm), the Herpes virus (cold sores and shingles), and infestation by parasitic organisms (scabies and head lice).
◆ **Inflammatory conditions**, known as dermatitis or eczema, cause itchiness, reddening and often oozing blisters. This may be a reaction to touching a substance (contact dermatitis), or a partly inherited sensitivity of the skin (particularly in children), as in atopic eczema.
◆ **Skin tumours** include benign (non-cancerous) growths such as warts and most moles, and skin cancer. The most dangerous skin cancer, melanoma, starts in the melanocytes – often in a mole. Less dangerous but more common are squamous cell carcinoma and basal cell carcinoma. All skin cancers are usually triggered by excessive unprotected exposure to sunlight.

Bones and skeleton

The numbers or star following the answers refer to information boxes on the right.

QUESTION NUMBER	ANSWERS
★ 95	**Thigh and ear**
110	**True** – other primates do have a bone
158	**Osteoarthritis** (disorder of joints, not bones) (7)
199	**True** (as strong as iron, but much lighter) (1)
242	**The hyoid** (in the throat) (5)
263	**Joints** (4)
301	**Rib** (2)
302	**Thigh bone** (2)
303	**Kneecap** (2)
304	**Finger or toe bone** (plural: phalanges) (2)
305	**Spinal, or back, bone** (2)
344	**Cartilage** (1)
404	**A: 206** (2)
417	**'Dem [Dry] Bones'** – Ezekiel chapter 37
465	**Bone fractures** (3)
531	**Calcium** (1)
541	**NSAID** (non-steroidal anti-inflammatory drug) (7)
685	**True** (2)
696	***Hamlet*** – act V, scene 1
922	**Bursitis** (7)
923	**Osteoporosis** (6)
934	**False** (some joints are fixed) (4)

Core facts (1)

◆ Bones are composed of mineral salts – mainly **calcium** – and a network of fibres made of the protein **collagen**. Most bones are dense near the outside and light and porous in the middle.

◆ The core of many bones contains **marrow** – a soft tissue where red and white blood cells are formed and passed into the veins.

◆ Babies' bones are made largely of softer **cartilage**, and harden as they absorb calcium.

◆ The **skeleton** is the internal 'scaffolding' of the human body, giving the body its basic shape and protecting vital organs.

Skull

Mandible (jawbone)

Spine: 7 cervical (neck) vertebrae; 12 thoracic (chest) vertebrae

Humerus (upper arm bone)

Sternum (breastbone)

Ribcage: 24 costals (ribs)

Spine: 5 lumbar (back) vertebrae; ends with sacrum and coccyx behind the pelvis

Radius

Ulna

Pelvis

Each hand: 8 carpals (wrist bones); **5 metacarpals** (main hand bones); **14 phalanges** (finger bones)

Femur (thigh bone)

Patella (kneecap)

Tibia (main shin bone)

Fibula

Calcaneus (heel bone)

Each foot: 7 tarsals (ankle bones); **5 metatarsals** (main foot bones); **14 phalanges** (toe bones)

The skeleton (2)

A normal adult has **206 separate bones**, but a newborn child has at least 300, many of which later fuse together – for example, the sacrum, near the base of the adult spine, is a single bone fused from five vertebrae. Anatomists divide bone structure into the **axial skeleton** (the 80 bones of the skull, spine and ribcage) and the **appendicular skeleton** (126 bones, comprising the limbs and the pelvic and pectoral girdles which connect the limbs to the spine).

Fractures (3)

In a **simple** (closed) fracture, the bone does not protrude through the skin; in a **compound** (open) fracture, it does. In addition, doctors distinguish several types of fracture by the shape of the break or other damage.

TRANSVERSE FRACTURE A horizontal break running straight across the bone.

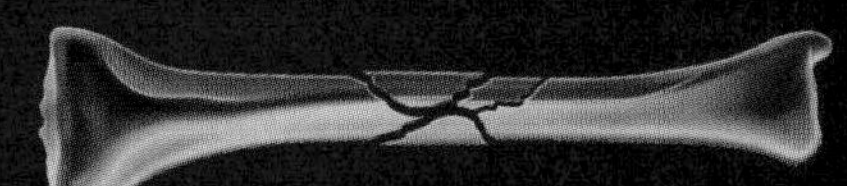

COMMINUTED FRACTURE A fracture with more than two bone fragments.

SPIRAL (OBLIQUE) FRACTURE A slanted break caused by twisting the bone.

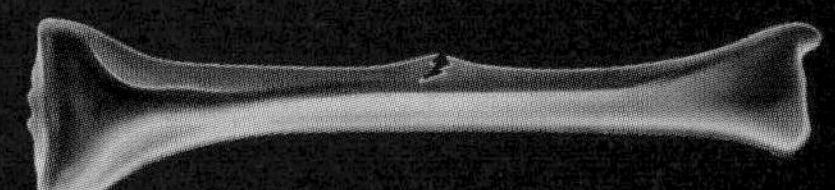

UNDISPLACED FRACTURE A break in which the bone fragments stay in place.

The joints

4

Bones are joined in different ways. They may be **fixed**, with no movement (for example, the bones of the skull, joined by cartilage); **semi-movable**, with some flexibility (for example, in the spine); or **fully movable** (synovial), of which there are six types:

- **Hinge** Allows movement in one plane only, as in the elbow, knee and fingers.
- **Pivot** A projection in a socket allowing only turning, as in the neck vertebrae.
- **Ball-and-socket** Sliding ball in socket allows movement in all directions, as at the shoulder.
- **Gliding** Flat bone surfaces slide over each other, as in the wrist bones.
- **Ellipsoidal** Egg shape in a cavity allows some movement in most directions, as at the base of the fingers.
- **Saddle** Allows limited movement in two planes, as at the base of the thumb.

SWINGING JOINT The thigh and shin bones meet at a hinge joint (left), which moves in one plane only.

FREE ROTATION The hip joint (right) has a highly flexible ball-and-socket joint.

WEIRD AND WONDERFUL

5

The **hyoid bone**, which sits underneath the back of the jaw, is a 'floating' bone – it has no direct attachment to the skeleton. Its supporting structures include the tongue and larynx muscles, which are used for speech.

95

Big and small

The biggest human bone is the thigh bone, or **femur**. The smallest bone is the **stapes** ('stirrup') in the middle ear, one of three tiny ossicles, an arrangement of bones that transmit sound. It measures approximately 3 mm (1/8 in) in length.

Brittle bones

6

As the human body ages, it becomes less efficient at absorbing calcium. Nerves and muscles need calcium to function – if levels in the bloodstream are low, it is taken from the bones, making them thin and porous. Between the ages of 40 and 70, bones may lose 30 per cent of their weight, making them brittle and more liable to break.

This condition, known as **osteoporosis**, affects two and a half times as many women as men. The female sex hormone, oestrogen, helps to maintain bone mass, but levels fall with age. Sufferers may become stooped as spinal bones are compressed – a feature known as 'dowager's hump'.

WEAK BONE This thigh fracture shows the spongy holes and brittleness associated with osteoporosis.

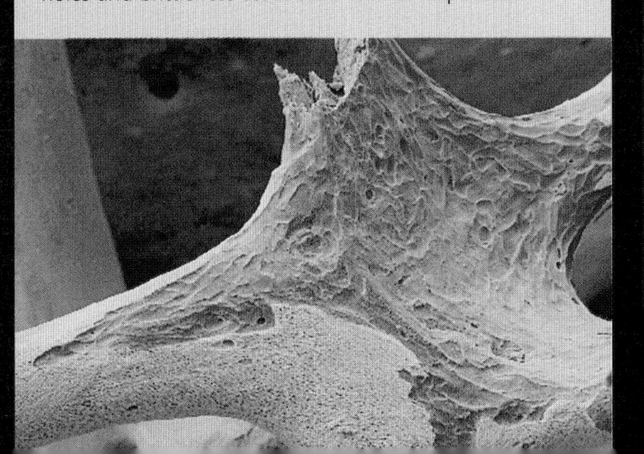

Other diseases of joints and bones

7

- **Joint disorders** cause pain and stiffness. Osteoarthritis is caused by loss of cartilage between the bones; rheumatoid arthritis is characterised by severe inflammation around the joints. 'Housemaid's knee' is a form of bursitis – inflammation of a bursa, a fluid-filled sac around a joint. A slipped (or prolapsed) disc occurs when one of the pads of cartilage between the vertebrae protrudes and presses on a nerve.
- **Infection** of bone is known as osteomyelitis, and can be contracted from bacteria in an open wound or an infection elsewhere in the body.
- **Metabolic disorders** affect bone composition. Calcium imbalance is a symptom of Paget's disease, and lack of vitamin D (needed to process calcium) causes rickets – both conditions lead to bone-softening.
- **Tumours** can be cancerous or non-cancerous. Cancerous tumours may originate in bone (primary tumours) or, more commonly, spread into the bone from other parts of the body (secondary tumours).

QUESTION NUMBER

The numbers or star following the answers refer to information boxes on the right.

ANSWERS

69	**C: Smooth muscle** 1 6
89	**Repetitive strain injury** 7
96	**Buttocks and ear** 5
166	**Muscles and tendons** – also in joints
192	**False** (they can only pull) 1 3
259	**B: Join muscles to bones** 1
306	**Calf muscle** 2
382	**True** (but nerve signals influence its speed) 1 6
523	**D: Types of muscle filament** 4
569	**Front of the upper arm** 2
570	**The buttocks** 2 5
692	**Popeye** – first appeared 1929
718	**B: The deltoid** 2
834	**B: 35 per cent** 5
924	**Tendinitis** (or tenosynovitis) 7
★ 970	**True**

Muscles and movement

Core facts 

◆ Muscles have the ability to **contract** (shorten) and exert a pull on bones or on other muscles or tissues. They cannot push.

◆ There are three types of muscle: **skeletal muscle**, usually under conscious control; **smooth muscle**, in many internal organs, which is controlled unconsciously; and **cardiac muscle**, in the heart, which can contract rhythmically of its own accord.

◆ **Tendons** attach skeletal muscles to bones. To produce body movements, muscles, bones and tendons act together as **levers**, with joints forming the pivots.

Anterior view **Posterior view** 2

Frontalis (forehead muscle)
Orbicularis oculi (eyelid and orbit)
Deltoid
Pectoralis major
Biceps brachii
Triceps brachii
Latissimus dorsi
Rectus abdominis ('six pack')
Gluteus medius
Gluteus maximus
Sartorius
Hamstrings
Semimembranosus
Semitendinosus
Biceps femoris
Quadriceps
Vastus medialis
Rectus femoris
Vastus lateralis
Gastrocnemius
Peroneus longus
Tibialis anterior
Extensor digitorum longus

Muscle machines 

The muscular system converts chemical energy manufactured in the body's cells into mechanical energy, which moves the bones and tissues. Muscles create movement by **contracting**. To move the skeleton, a contraction is transmitted to a bone just beyond the joint, where the muscle is attached by a tendon – a band of fibrous tissue. The contraction bends, straightens or rotates the joint.

Muscles, tendons, bones and joints can be thought of as **lever systems** that magnify small movements, with the joint acting as the fulcrum, or pivot.

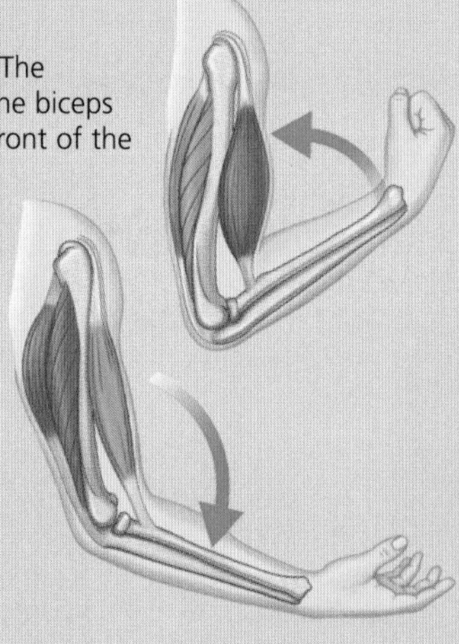

LEVERING UP The contraction of the biceps muscle on the front of the upper arm (top) bends the arm. The contraction of the triceps muscle at the back of the upper arm staightens the arm.

Contraction 4

Skeletal muscles consist of hair-thin fibres made up of **myofilaments** – tiny strands of protein. The protein filaments are arranged in alternate bundles of thin strands (made mainly of **actin**) and thick strands (made of **myosin**), which overlap one another at each end.

Myosin filaments have minute projections called **cross-bridges**. When stimulated by a nerve, these are thought to act like a ratchet, pulling the thin actin filaments on either side and making them slide between the myosin filaments, shortening the muscle.

RAW POWER Four-times world champion strong-man, Magnus Ver Magnusson pulls a 30-tonne truck.

Muscle statistics ❺

◆ **Average percentage of body weight** Men 45 per cent; women 35 per cent.
◆ **Biggest** Gluteus maximus, in the buttock, weighing about 1 kg (2½ lb).
◆ **Smallest** Stapedius, in the middle ear, responsible for damping loud sound); about 5 mm (1/5 in) long.
◆ **Longest** Sartorius, from outer hip to inner knee; about 50 cm (20 in) long.
◆ **Longest group** Erector spinae, along the spine; about 90 cm (35 in) long.

970

Body building

The body uses proteins from food to maintain and repair muscle tissue. Body builders combine intense **exercise** with a **high-protein diet** to increase the mass of skeletal muscles and decrease body fat. Protein-building **anabolic steroids** (synthetic male hormones) can help to increase muscle mass, but can have side effects and their use in sport is illegal in many countries.

BODY BUILDER Weight-bearing exercise increases muscle strength and, to some extent, bulk.

Muscle types ❻

The human body contains three types of muscle, each with a different appearance and function:

SUPPORTING THE BONES Skeletal muscle is made up of overlapping fibres which appear as dark and light bands called striations.

◆ **Skeletal muscle** makes up the majority of the body's muscular system. It is attached to the bones and enables the skeleton to move. It is also known as **voluntary** muscle, because it operates under conscious control.
◆ **Smooth muscle** consists of short, fine fibres with no striations, and is found in the walls of internal systems such as the digestive, circulatory and reproductive systems. It can contract for prolonged periods. Smooth muscle is also known as **involuntary** muscle, because it is controlled by the autonomic nervous system, which operates outside of conscious control.
◆ **Cardiac muscle** is found in the heart. It has short, branching, interlinked fibres forming a network that transmits rhythmic waves of contraction. It enables the heart to beat continually, without tiring. Autonomic nerve signals speed up and slow down the heartbeat, but do not initiate the contractions.

Muscle disorders ❼

◆ **Muscular atrophy** occurs when nerve signals that stimulate muscles fail, causing paralysis and wasting. It is a symptom of motor neurone disease (amyotrophic lateral sclerosis – AMS) and the viral disease poliomyelitis.
◆ **Myopathy** is characterised by wasting and weakness unconnected with the nervous system. Examples include polymyositis, which causes inflammation and tenderness, and muscular dystrophy, in which muscles grow weaker and lose their ability to contract.
◆ **Injury and strains** are often caused by high levels of activity. For example, repetitive strain injury (RSI) causes pain in muscles and tendons that have been subjected to rapid, repeated movements. Inflammation of a tendon – the cause of conditions such as tennis and golfer's elbow – is known as tendinitis.

Heart and circulation

QUESTION NUMBER

The numbers or star following the answers refer to information boxes on the right.

ANSWERS

- **8** Arteries ❶ ❺
- ★ **131** D: 0.001mm
- **133** B: 2.7 million litres (600 000 gallons) ❻
- **270** Four ❷
- **374** Christiaan Barnard ❽
- **376** William Harvey ❽
- **411** Heart (by Douglass Cross and George Cory) ❹
- **416** Joseph Conrad – published 1902
- **418** The strings of my heart (by James Hanley) ❹
- **497** True ❼
- **548** ACE inhibitor – angiotensin converting enzyme
- **613** The heartbeat ❶ ❸
- **627** Arrhythmia ❼
- **649** B: 1967 ❽
- **663** Heart transplant ❽
- **720** D: Coronary thrombosis ❼
- **725** The heart – from Greek *kardia*, heart
- **763** Aorta ❺
- **769** Healthy heart ❷
- **788** In the legs ❺ ❻
- **804** Heart attack – caused by atherosclerosis/thrombosis
- **967** True ❼

Core facts 

◆ The **circulatory system** is the body's main transport network. Its function is to move blood in a continuous circuit around the body.

◆ The circulation is driven by the **heart**, which pumps blood through the lungs to collect **oxygen**. It then pumps the oxygenated blood through the rest of the body.

◆ Blood vessels carrying blood away from the heart are called **arteries**, those returning it to the heart are **veins**. Arteries and veins are linked by tiny vessels called capillaries.

◆ Surgical advances in the 20th century have allowed repair and maintenance of the heart through technology such as **pacemakers**.

The heart 

The circulation is driven by the heart, a four-chambered muscular pump in the middle of the chest. The **pulmonary circulation** on the right side takes blood through the lungs to collect **oxygen**, while the chambers on the left pump oxygen-rich blood through the **systemic circulation** – the rest of the body. The left side has stronger muscles, so the heartbeat is felt mostly on the left. The four chambers (the atria and ventricles) are sealed inside a fibrous sac – the **pericardium**.

LIFE SOURCE Muscles wind around the heart in a spiral, creating a wringing action from top to bottom as they contract.

Right atrium receives returning blood from venae cavae and passes it to right ventricle.

Inferior vena cava returns blood to heart from the lower body.

Superior vena cava returns blood to the heart from the upper body.

Aorta takes blood from left ventricle to all organs, except the lungs.

Left pulmonary artery takes deoxygenated blood from right ventricle to left lung.

Left pulmonary vein brings oxygen-rich blood to left atrium from left lung.

Left atrium receives blood from pulmonary veins and passes it to the left ventricle.

One-way valves between the atria and ventricles, and at the exit of the ventricles, prevent the backflow of blood.

Left ventricle pumps oxygenated blood into the aorta and round the rest of the body.

Right ventricle pumps deoxygenated blood into the pulmonary arteries.

Pacemakers ❸

A centre in the wall of the right atrium called the **sinoatrial (SA) node** is the heart's natural 'pacemaker'. It generates regular electrical impulses 60 to 80 times a minute when a person is resting, speeding up during physical activity. The impulse spreads to the heart muscle via a second node, the **atrioventricular (AV) node**, causing the atria and ventricles to contract and pump the blood. If the SA or AV node does not work properly, an **electronic pacemaker** can be implanted to supply timed electric impulses to maintain the heartbeat.

Heart symbolism ❹

The ancient **Egyptians** saw the heart as the centre of intelligence and emotion. In **Chinese** tradition, it is the origin of happiness. The **Romans** associated it with love – Cupid, their god of love, was a winged boy firing arrows through people's hearts. The heart is still associated with love, bravery and sadness, perhaps because its response to hormones associated with emotions are immediately felt. The heart does not cause emotions, but it does reflect them.

Circulatory system

❺

Arteries carry red oxygenated blood at high pressure away from the heart, with the exception of the pulmonary arteries, which carry deoxygenated blood to the lungs. All arteries have thick, muscular, elastic walls. **Veins** carry deoxygenated blood (shown in blue) at low pressure back to the heart, with the exception of the pulmonary veins, which bring oxygenated blood to the heart. Veins have thinner, less elastic walls than arteries, and often contain one-way valves to prevent backflow. Arteries and veins are connected by tiny **capillaries**, with walls one cell thick.

Jugular vein drains the head.

Superior vena cava drains the upper body and arm.

Aorta supplies oxygenated blood to the body.

Pulmonary vein picks up oxygenated blood from the lung.

Pulmonary artery supplies deoxygenated blood to the lung.

Inferior vena cava drains the lower body.

Renal vein drains the kidney.

Renal artery supplies oxygenated blood to the kidney.

Common iliac artery supplies oxygenated blood to the leg.

Common iliac vein drains the leg.

Great saphenous vein drains the leg.

★ 131

Big and small

The largest blood vessel, the **aorta**, is 2.5-3 cm (1-1 1/5 in) in diameter and about 45 cm (18 in) long. It supplies oxygenated blood to most of the body's organs. The smallest blood vessels, **capillaries**, are less than 0.001 mm (1/2500 in) in diameter – just wide enough for a red blood cell to pass through.

Facts and figures

❻

◆ **Weight** The human heart weighs between 250 and 400 g (9 and 14 oz).

◆ **Heartbeats** The heart beats 70 times a minute, on average – 100 000 times a day, 36.5 million beats a year.

◆ **Circulation** When the body is at rest, each heartbeat circulates about 75 ml (2 fl oz) of blood. All of the blood in the body's blood vessels is circulated about once a minute – 7500 litres (1650 gallons) a day, 2.7 million litres (600 000 gallons) a year.

◆ **Volume** Blood accounts for about 8 per cent of total body weight.

◆ **Longest veins** The great saphenous veins, from the ankles to the hips.

◆ **Circulatory system** All of the body's blood vessels laid end to end would stretch more than 96 600 km (60 000 miles) – almost enough to wind two and a half times around the Earth.

FAST FLOW Blood in the aorta (background) travels at about 38 cm per second (15 in/sec).

Major disorders

❼

Cardiovascular disease (including heart disease and stroke) is responsible for around 30 per cent of deaths worldwide.

Disorder	Features
Aneurysm	Swelling due to weakness in artery wall.
Arrhythmia	Irregular or abnormally fast/slow heartbeat.
Atherosclerosis	Build up of fatty deposits on artery walls, narrowing the blood vessels.
Congenital heart disease	Heart defects present at birth, such as a valve deformity.
Hypertension	Raised blood pressure, causing heart and blood vessel damage.
Thrombosis	Blockage of vessel by a blood clot.

HEART SCAN This ultrasound shows the healthy heart of a 30-year-old man.

TIMESCALE

❽

▶ **1628** English doctor William Harvey publishes the principles of blood circulation.

▶ ***c.*1730** English physiologist Stephen Hales first measures blood pressure (of a horse).

▶ **1896** Italian doctor Scipione Riva-Rocci invents an efficient, non-invasive tool for measuring blood pressure – the sphygmomanometer.

▶ **1912** US doctor James B. Herrick identifies the features of coronary thrombosis.

▶ **1952** First artificial heart valve implanted.

▶ **1953** First successful use of heart-lung machine during open-heart surgery.

▶ **1960** American electrical engineer Wilson Greatbatch patents the first implantable artificial pacemaker.

▶ **1967** South African surgeon Christiaan Barnard performs the first successful human heart transplant.

▶ **1977** Angioplasty first used to widen blocked arteries.

QUESTION NUMBER

The numbers or star following the answers refer to information boxes on the right.

ANSWERS

34 **Vampire bats** – from Central and South America

36 **Rhesus** 4

42 **Eight** (A, B, AB, O, and in each case Rh+ and Rh–) 4

64 **D: Septicaemia** (the others are white cell disorders) 8

★ 80 **Leeches**

98 **Bone marrow** 5

266 **5 litres (9 pints)** 1

267 **Iron** 3

293 **Red Cross** – founded 1864

297 **Red** 2 3

313 **Tony Hancock** 4

345 **Corpuscles** 1

405 **B: 5 million** 2

490 **Nelson's blood** – after victor of Trafalgar (1805)

614 **Vampires** – from Bram Stoker's *Dracula* (1897)

661 **Transfusion** 6

782 **Haemoglobin** 3

812 **George Patton** – 1885-1945

832 **D: 90 per cent** 2

855 **True** – men, 5.2 million per mm³; women, 4.6 million

892 **Blood plasma** 1 2

988 **Michael Flanders and Donald Swann** – 'The Hippopotamus Song'

Blood

Core facts 1

◆ Blood consists of three types of cell or corpuscle – **red blood cells, white blood cells** and **platelets**. They are suspended in a fluid called **plasma**.

◆ An average adult circulatory system contains about **5 litres** (9 pints) of blood; the system of a 40 kg (6 stone) child contains half that.

◆ Blood carries **oxygen**, nutrients and water to the tissues. It carries away **carbon dioxide** and other wastes and distributes the **hormones** that regulate bodily functions.

◆ Blood also helps to maintain a steady body **temperature**, fights **infection** and operates **clotting** mechanisms to limit bleeding.

What is blood? 2

◆ Separated into its components, blood consists of about 45 per cent **cells** and 55 per cent **plasma** – a liquid made of 90 per cent water and 10 per cent other substances such as glucose and hormones.

◆ Disc-shaped **red blood cells** (erythrocytes) carry oxygen and carbon dioxide around the body. A 1 mm³ drop of blood contains 4.5 million to 5 million red blood cells.

◆ **White blood cells** (leucocytes) control immunity and clean up debris. A 1 mm³ drop contains 4000 to 10 000.

◆ Fragments of cells called **platelets** (thrombocytes) help in clotting blood – a 1 mm³ drop contains 250 000 to 500 000.

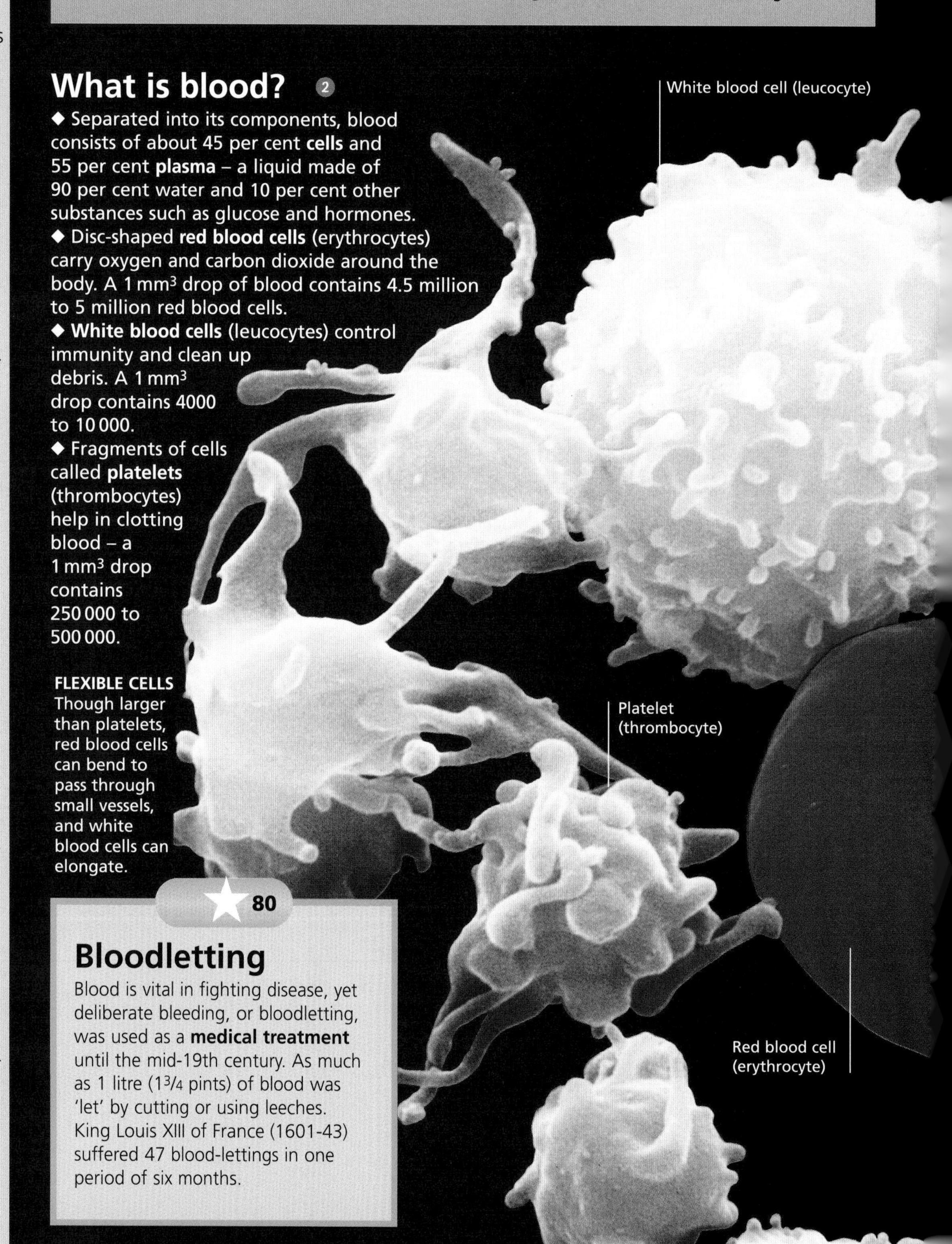

FLEXIBLE CELLS
Though larger than platelets, red blood cells can bend to pass through small vessels, and white blood cells can elongate.

★ 80

Bloodletting

Blood is vital in fighting disease, yet deliberate bleeding, or bloodletting, was used as a **medical treatment** until the mid-19th century. As much as 1 litre (1¾ pints) of blood was 'let' by cutting or using leeches. King Louis XIII of France (1601-43) suffered 47 blood-lettings in one period of six months.

Red cells

These take their colour from **haemoglobin**, a substance which contains most of the body's iron. Haemoglobin molecules link up with oxygen in the lungs to form bright red oxyhaemoglobin. In the body's tissues, the oxygen is used to 'burn' nutrients, releasing energy for bodily processes. The haemoglobin then links up with carbon dioxide to make dull red carbaminohaemoglobin, and the carbon dioxide is released back into the lungs. This process is known as **internal respiration**.

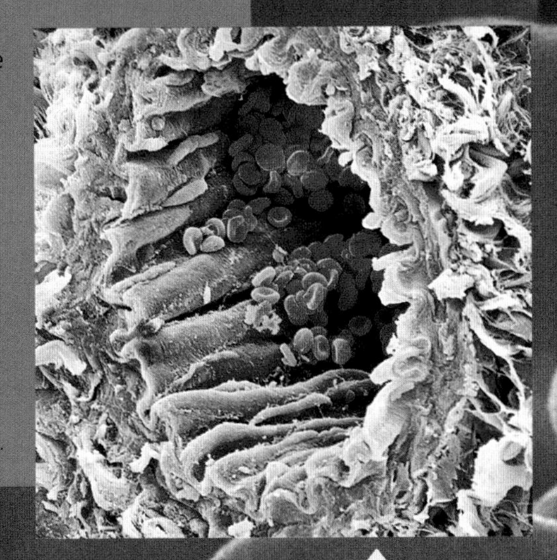

SPECIAL DELIVERY Red blood cells course through an artery, carrying oxygenated blood to the body tissues.

Blood groups

In 1901, Austrian scientist Karl Landsteiner identified four types of blood – **O, A, B** and **AB** – based on differences in the proteins attached to the red cells. Forty years on, he discovered the positive and negative **Rhesus system**, first observed in rhesus monkeys. His work is vital to transfusions – matching groups avoids a dangerous immune response.

Life cycle

Bone marrow contains cells called **haemocytoblasts**, which develop into blood cells and platelets. Red cells last for 100 to 120 days, and are then destroyed in the liver and spleen and their components re-used. White cells, which are made in bone marrow but mature in lymphoid tissue, last from a few hours to years. Platelets last just ten days.

WEIRD AND WONDERFUL

In 1667, French surgeon Jean-Baptiste Denys carried out the first documented **blood transfusion** into a human being – from a sheep. The recipient, a feverish boy weakened by several bloodlettings, recovered.

HEALING MESH Strands of fibrin trap blood cells in an open wound.

Clotting

When a blood vessel is broken, **platelets** stick to the break and to each other to form a plug. They also stimulate the blood to make **fibrin** threads, which form a mesh to trap blood cells. The resulting blood clot dries to form a scab. Under the scab, the skin edges grow towards each other until they meet and heal, then the scab falls away.

Major disorders

◆ **Anaemia** is a low red blood cell count or low haemoglobin levels. It reduces the blood's oxygen-carrying capacity. Causes include poor diet, infection or genetic conditions such as thalassaemia.

◆ **Bleeding disorders** include genetic diseases such as thrombocytopaenia and haemophilia. In these diseases, a low number of platelets reduce the clotting abilities of the blood, which can lead to excessive bleeding.

◆ **White cell disorders** include leukaemia, in which uncontrolled production of abnormal white cells occurs; and leucopaenia, very low numbers of white cells.

◆ **Infections** such as malaria, glandular fever and HIV attack blood cells. In blood poisoning (septicaemia) bacteria multiply in the bloodstream, causing high temperature and joint and muscle pain.

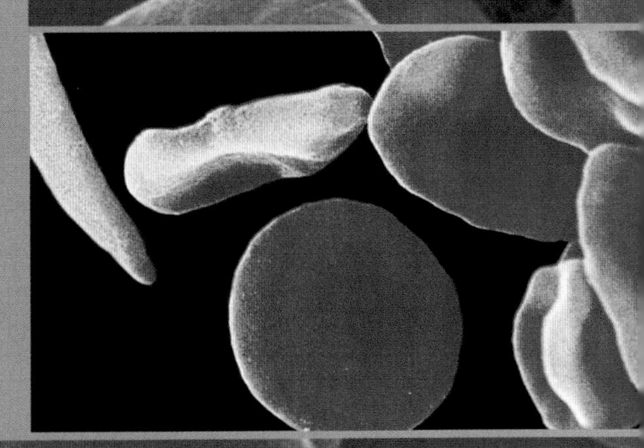

GENETIC ANAEMIA The inherited disease sickle-cell anaemia causes crescent or sickle-shaped red blood cells which increase blood viscosity.

Lymph and immune system

QUESTION NUMBER

The numbers or star following the answers refer to information boxes on the right.

ANSWERS

- **9** Allergy ❽
- **67** **D: A worm** – nematode worms block lymph vessels
- **243** Lymphocytes ❸
- **321** **Hodgkin's disease** (Thomas Hodgkin, 1798-1866) ❶ ❷
- **393** Macrophages ❸
- **466** Lymph nodes ❶ ❷
- **546** Antihistamine ❽
- **671** Immune system ❶
- **672** Lymphatic system ❶ ❷
- **711** B: Antigens ❹ ❺ ❻
- **717** C: 1 million ❹
- **752** Antibodies ❹
- **753** Antigen ❹
- **823** Autoimmune disease ❽
- ★ **872** Cowpox
- **901** Vaccination ❶ ❻ ★

Core facts ❶

◆ Straw-coloured **interstitial fluid**, similar to blood plasma, bathes all body tissues. A one-way system of tubes called **lymphatics** drains this fluid back into the bloodstream.

◆ **Lymph nodes** (or 'glands') and spongy tissue in the adenoids, spleen, thymus and tonsils filter out disease organisms and waste.

◆ The lymphatic system works with the immune system to fight disease. Its **white blood cells** destroy potentially damaging invaders such as bacteria, cancer cells and viruses.

◆ The disease-preventing actions of white blood cells in lymph form the basis of **vaccination**.

Lymphatic system ❷

Body tissues are bathed in interstitial fluid, which passes out of blood capillaries into the spaces between cells and mixes with cellular waste products. The fluid drains into a network of vessels, the lymphatic system, where it is known as **lymph**.

When the body moves, muscles push against the vessels and lymph passes through in one direction – valves prevent backflow. It filters through **nodes** of spongy tissue and organs such as the spleen where **white blood cells** remove disease organisms, then drains into the subclavian veins at the collarbone.

White blood cells ❸

The white blood cells in the lymphatic system maintain immunity. **Lymphocytes**, known as B cells and T cells, fight infection. **Neutrophils** and huge scavenger cells called **macrophages** engulf and digest (phagocytise) bacteria and debris. **Basophils** and **eosinophils** control inflammation.

Antibody immune response ❹

Bacterial infections stimulate B cell lymphocytes to activate **antibody-mediated immunity**. The B cells react to foreign proteins (antigens) attached to the surface of bacteria and begin to multiply and produce antibodies (also proteins). Antibodies bind to antigens, then bind with phagocytic cells which engulf and destroy both antibody and antigen.

WELL ANCHORED Ridges and projections on the surface of B cells help them bind to antigens.

Cellular immune response ❺

When cancers or organisms such as viruses or parasites enter the body's cells, they go beyond the reach of the B cell antibodies circulating between the cells, so T cells take over the immune response. T cells engulf phagocytic cells which have consumed foreign material – a process known as **cell-mediated immunity**. When phagocytes consume molecules of an invading disease, antigens appear on their surface and are identified by T cells with matching receptors. The T cells multiply, producing **cytotoxic ('killer') T cells** which lock onto the antigens and destroy the cells.

T cells also play a role in antibody-mediated immunity – **helper** and **suppressor T cells** inhibit or promote the actions of B cells.

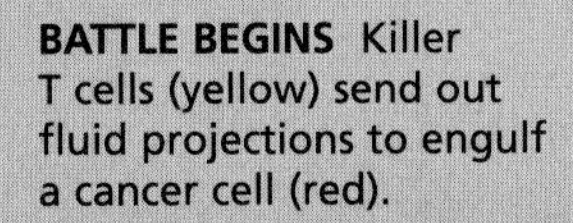

BATTLE BEGINS Killer T cells (yellow) send out fluid projections to engulf a cancer cell (red).

Memory cells ❻

During the body's first exposure to a particular antigen, B cells and T cells produce **memory cells** which keep a record of the antigen. When these cells meet the same antigen again, they divide and produce large amounts of antibodies to prevent disease from taking hold. Some memory cells, such as those for the **common cold** virus, are short-lived, but most last for many years and some last a lifetime – the basis of vaccination.

SOURCE OF IMMUNITY White blood cells (shown in blue) originate in the bone marrow – here magnified 600 times – then migrate to the thymus to mature.

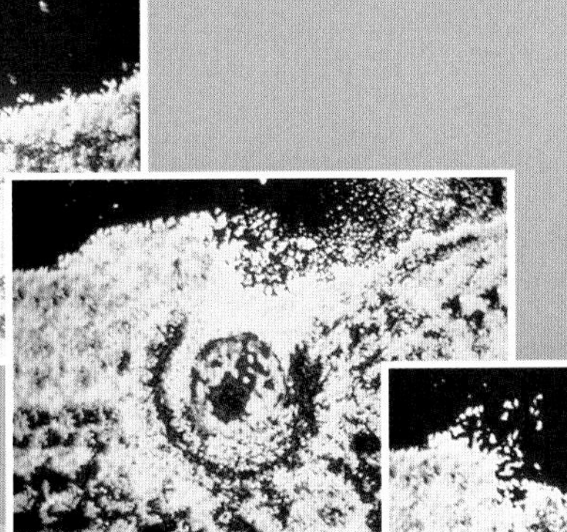

SILENT INVASION When HIV (orange) enters a T cell (white) it multiplies rapidly, but years can pass before its effects are felt.

Immune deficiency ❼

Many things can weaken the body's ability to fight infection, from poor diet and stress to chemotherapy and drugs such as steroids. Surgical removal of the spleen, which makes and stores lymphocytes, has the same effect.

Weakened immunity is a characteristic of **HIV** (the **human immunodeficiency virus**). It invades macrophages and the T helper cells which promote antibody production. The number of these cells gradually falls, leaving the body prone to a wide range of infections.

★ 872

Vaccination ★

In 1796, British physician **Edward Jenner** carried out the first vaccine experiment. Jenner knew that a **cowpox** (*vaccinia*) infection prevented smallpox, and deliberately exposed patients to a mild dose of cowpox to create a resistance to smallpox.

Autoimmunity ❽

The antibody response does not always bring benefits. Sometimes, for unknown reasons, B cells attack 'self-antigens' – proteins in the body's own tissues. This causes **autoimmune diseases** such as lupus erythematosus, myasthenia gravis and rheumatoid arthritis. B cells also attack **transplanted organs**, which they recognise as foreign material.

An incorrect immune response causes mild conditions such as **allergies**. Initial exposure to a harmless substance such as pollen can sensitise the immune system. Later exposures cause antibodies to attack the body's mast cells – found in areas such as the skin and nasal lining. The mast cells then release **histamine**, a substance which causes an allergic reaction.

QUESTION NUMBER

The numbers or star following the answers refer to information boxes on the right.

ANSWERS

137 A: 0.5 litres (1/10 gallon) 6

155 Pleurisy (affects pleura, not lungs themselves) 2 8

204 Pneumoconiosis, or silicosis 8

269 Heart and lungs 1 2

307 Windpipe 1 2

421 Pulmonary – from Latin *pulmo*, lung

472 Reflex 5

513 Carbon dioxide 1 4 6

542 Bronchodilator (dilates – or opens – airways) 8

728 Whooping cough 8

765 Lung 2

767 Bronchus 2

784 The intercostal muscles (or rib muscles) 3

787 Carbon dioxide 6

★ 813 Krebs cycle

928 Bronchitis 8

946 *Breathless* – starring Richard Gere

987 The Police – released 1983

994 Tuberculosis 8

1000 Pneumonia 8

Lungs and respiration

Core facts 1

◆ The **lungs** fill most of the chest (thorax), and are linked to the nose and mouth by the windpipe, or **trachea**.

◆ In **external respiration**, the lungs draw in air and absorb oxygen into the bloodstream. When oxygen reaches the body's tissues, it is used by cells to 'burn' or break down food materials and release energy in a process known as **cellular respiration**. The waste products, carbon dioxide and water, are taken to the lungs by the bloodstream and exhaled.

◆ The muscles which facilitate external respiration operate mainly under the control of the **autonomic nervous system**.

Anatomy of a lung 2

The **respiratory system** consists of the main windpipe (the trachea), tubes branching out from the trachea (the bronchi) and the **lungs**. The lungs are soft, spongy and cone-shaped, and are surrounded by a double-layered, fluid-filled membrane called the **pleura** which lubricates their movement. The left lung is smaller than the right, with two lobes rather than three, to allow space in the chest for the heart.

Pulmonary arteries and veins carry blood between the heart and the lungs. Blood entering the lungs picks up oxygen and circulates it, releasing it into the body tissues. The deoxygenated blood then flows back through the lungs to begin the cycle again.

★ 813 Cellular respiration

In a process called the **Krebs cycle**, cell mitochondria break down food molecules into energy-rich **adenosine triphosphate (ATP)**. Energy released from ATP fuels the key cellular processes, including growth, movement and reproduction.

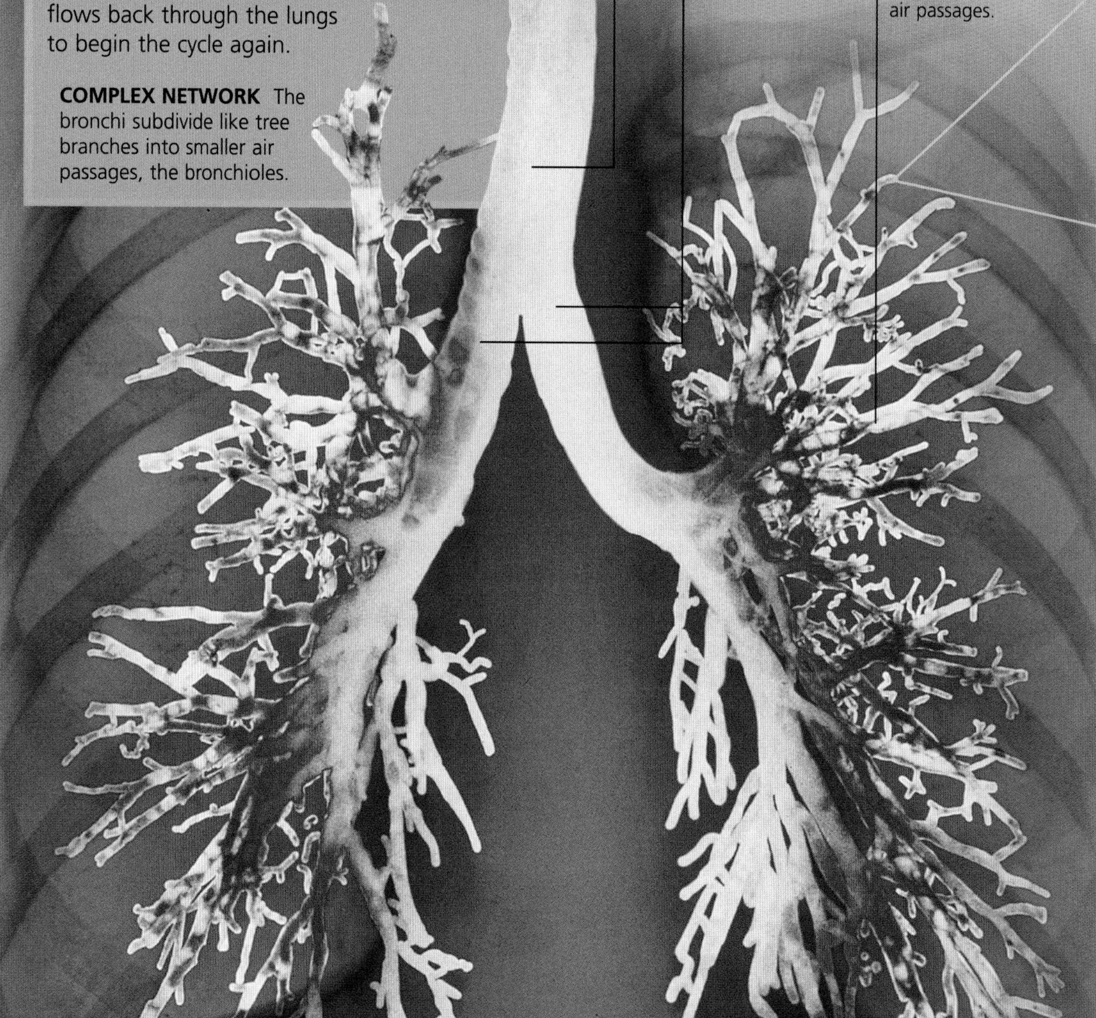

COMPLEX NETWORK The bronchi subdivide like tree branches into smaller air passages, the bronchioles.

Breathing

Breathing is powered by the **diaphragm** – a sheet of muscle separating the chest from the abdomen – and the **intercostal muscles** between the ribs.

These muscles work together to expand and contract the chest, sucking air in and then forcing it out.

Breathing is under the unconscious control of the autonomic (involuntary) nervous system, but it can be brought partially under conscious control.

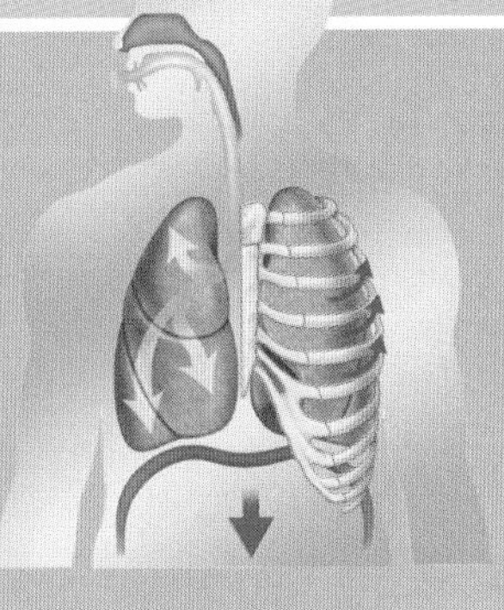

IN-BREATH The diaphragm tightens and flattens, and the intercostal muscles pull the ribs up and out. This increases the volume of the chest cavity and expands the lungs. Air flows in.

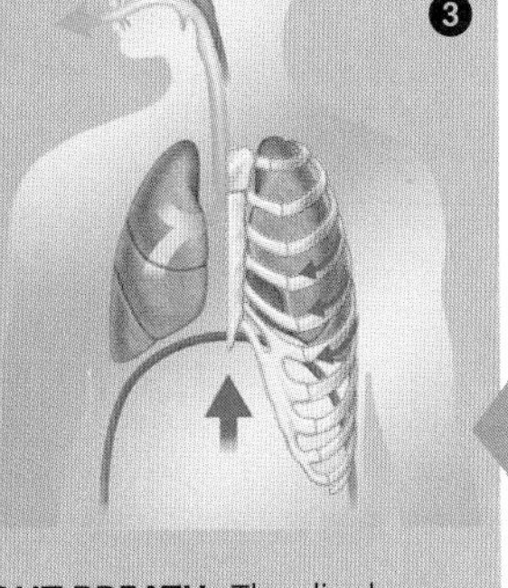

OUT-BREATH The diaphragm relaxes into a dome shape. The intercostal muscles relax, allowing the ribs to fall. Chest and lung volume decrease and air is pushed out.

WEIRD AND WONDERFUL

Estimates of the total **internal surface area** of the alveoli in an adult's lungs range from about 75 to 140 m² (800 to 1500 sq ft). This is as much as 70 times a person's total skin area, or up to half the area of a tennis court.

OXYGEN EXCHANGE The bronchioles end in clusters of air sacs (alveoli), where oxygen from air breathed in enters the bloodstream, and carbon dioxide from the blood enters the lungs to be breathed out.

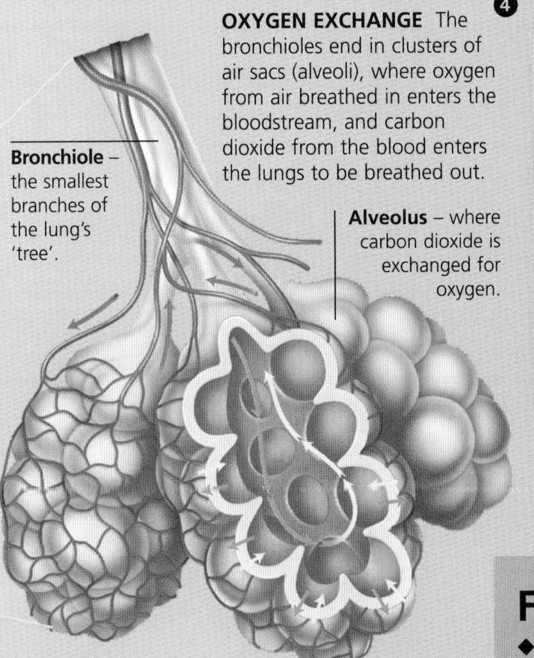

Bronchiole – the smallest branches of the lung's 'tree'.

Alveolus – where carbon dioxide is exchanged for oxygen.

Reflex action

Coughs and **sneezes** are reflexes. They are automatic muscular responses to nerve stimuli which rid the respiratory system of irritating objects. Sensory receptors in the trachea and nose send nerve signals to the brain to initiate the reflex.

Irritation of the digestive system or diaphragm causes **hiccups** – a spasm of the diaphragm with a sharp intake of breath and the closure of the vocal cords.

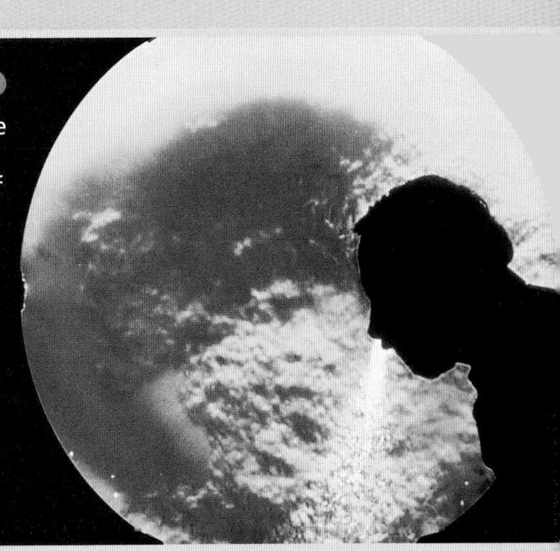

AT SPEED Sneezes can propel irritants up to 10 m (32 ft) at 60 km (37 miles) per hour.

CLEAN AIR The trachea and bronchi are lined with cilia (below) – hairs which act as a filter to keep out harmful microorganisms.

Facts & figures

◆ **Breathing** At rest, an average person takes 16 to 18 breaths per minute.

◆ **Air volume** On average, a resting adult takes in and expels 0.5 litres (1/10 gallon) of air in each breath, which can rise to 4.8 litres (1 gallon) with effort. A residual volume of about 1.2 litres (1/4 gallon) remains in the lungs.

◆ **Air composition** Inhaled air contains 21 per cent oxygen and 0.04 per cent carbon dioxide, exhaled air 16 per cent oxygen and 4 per cent carbon dioxide.

◆ **The lungs** The two lungs weigh about 1.2 kg (2½ lb) on average, and contain so much air that they float – which is why butchers call them 'lights'.

Lung disorders

◆ **Infections** The viral diseases influenza and pharyngitis can be serious if they extend to the bronchi (bronchitis) or lung tissues (pneumonia). Severe acute respiratory syndrome (SARS) is a type of pneumonia. Bacterial infections include tuberculosis and pertussis (whooping cough). Tuberculosis and pneumonia can cause pleural inflammation (pleurisy).

◆ **Obstructive/inflammatory diseases** Narrowing of the bronchi is a symptom of both asthma and chronic bronchitis. Severe and persistent obstructive conditions can lead to emphysema, in which the alveoli become distended.

◆ **Occupational diseases** Pneumoconiosis (coal miner's lung) – scarring of the bronchial tubes – results from long-term inhalation of dusts such as asbestos (causing asbestosis) and silica (silicosis).

QUESTION NUMBER

The numbers or star following the answers refer to information boxes on the right.

ANSWERS

23 **True** ❸ ❼

24 **False** (birds also have a larynx) ★

37 **Canines** ❺

111 **'Deep Throat'** – his identity was never revealed

157 **Pulpitis** (affects teeth, the others affect gums) ❻

194 **True** ★

257 **B: Orthodontics** ❼

264 **Teeth** ❹

★ 308 **Voice box**

403 **B: 20** ❺ ❼

441 **Dentist** ❼

612 **The *haka*** ❷

673 **Uvula** – visible at the back of the throat

674 **Pharynx** ❶ ❷

700 **Jane Horrocks** – also starred Michael Caine

761 **Larynx** ★

768 **Surface of the tongue** ❶ ❷

773 **False** ❶ ❹

852 **False** ❸

927 **Epiglottis** ❷

963 **True** ❻

Mouth, teeth and throat

Core facts ❶

◆ The mouth is where **digestion** begins, with the chewing of food. It also leads to the respiratory system, and modifies the sounds made by the voice box to produce **speech**.

◆ The throat (pharynx) is the crossover point of the digestive and respiratory systems. In humans, the mechanism of **swallowing** closes off the respiratory tubes to prevent food or drink from entering the windpipe.

◆ **Taste** is perceived through the reaction of sensory cells on the tongue in combination with the sense of smell.

◆ **Teeth**, embedded in the jawbones, are our hardest tissues, but cannot repair themselves.

The mouth and throat ❷

The process of **digestion** begins in the mouth and throat. The first stage is **mastication**, in which the teeth chew food to break it into smaller pieces and so increase the surface area available for nutrient absorption. Glands at the back of the mouth and underneath the tongue secrete **saliva** – up to 1.5 litres ($2\frac{1}{2}$ pints) a day – which lubricates the food and also contains enzymes that begin to digest carbohydrates.

The **tongue** is a mass of muscle fibres. It moves chewed food towards the back of the mouth ready for swallowing. On its surface, 10 000 taste buds contain specialised cells that react to different flavours. These work in combination with the sense of smell. During swallowing, the **epiglottis** (a flap of cartilage) covers the windpipe to prevent food from going down the 'wrong way'.

The throat, mouth and nasal cavity are also vital to **speech**. Their shape affects the tone of speech, and complex intonation is created by using the tongue, lips and teeth to modify basic sounds.

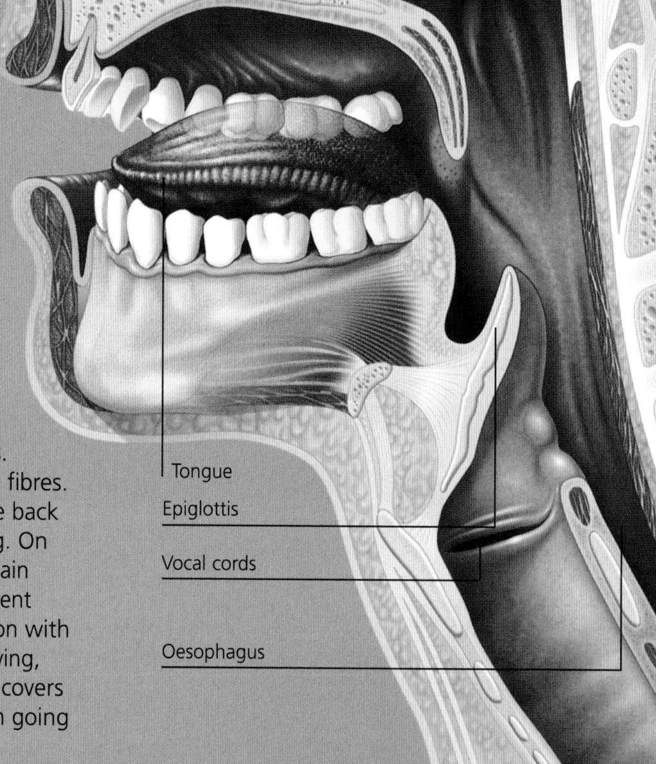

★ 308

The voice box ★

The **larynx**, or voice box, is a hollow tube near the top of the trachea. On either side of the larynx, the **vocal cords** (two small folds of tissue) form the **glottis** – a slit in which air vibrates to produce sounds. The size of the larynx governs the length of the cords and thereby the **pitch** of the voice. Women and children have a smaller larynx and shorter cords than men, causing a higher pitch.

WEIRD AND WONDERFUL ❸

Most people have **two sets of teeth** in life – milk teeth as a child, followed by one set of adult teeth. But sometimes a third set grows through later in life, and in 1896, a woman in France reportedly grew a fourth set.

Tooth structure

The **crown** of a tooth is the area that protrudes beyond the gumline; the **root** is hidden. The narrowed part where the tooth meets the gumline is the **neck**.

Enamel The hard, non-living coating of the crown.

Gingiva (gum) Protective tissue which helps to anchor the tooth firmly in the jaw.

Cementum A bone-like substance forming the tough coating of the root.

Periodontal ligaments Small fibres between the root and jawbone which anchor the tooth and absorb shocks.

Blood vessels Circulatory supply to the pulp.

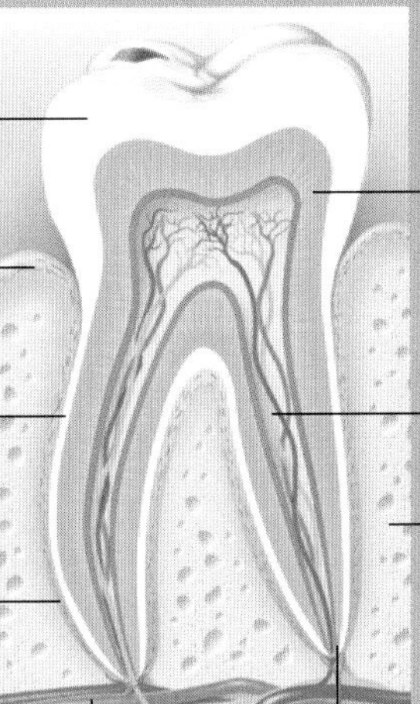

Dentine The tough, ivory-like body of a tooth, made mostly of minerals; contains some living cells.

Pulp The soft, inner core of a tooth, made of connective tissue, blood vessels and nerves; also called the 'nerve'; extends into the root canal.

Jawbone Deep sockets of bone (alveoli) encase the roots.

Apical foramen The hole at the base of each root through which nerves and blood vessels pass.

Permanent teeth

Adults have twelve molars, eight premolars, four canines and eight incisors. **Molars** are grinding teeth with four cusps (peaks). **Premolars** are broader grinding teeth, usually with two cusps (missing in children). **Canines** are sharp, pointed tearing teeth. **Incisors** are chisel-shaped cutting teeth.

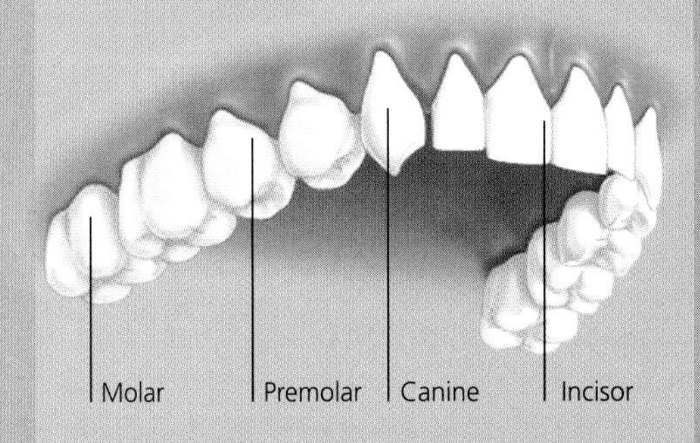

Mouth and throat disorders

Tooth decay begins as a small cavity in the enamel and spreads through the tooth, sometimes to the pulp (pulpitis), and can cause an abcess if the root becomes infected.
Tooth growth disorders include impacted teeth (one tooth colliding with another as it grows) and malocclusion (failure of the upper and lower teeth to come together properly when biting).
Periodontal problems affect the gums, and include gingivitis (inflammation) and periodontitis (long-term inflammation, causing loose teeth). The usual cause of gum problems is a build-up of plaque from poor dental hygiene.
Oral cancer causes tumours in the mouth. The risk is increased by smoking, chewing tobacco or drinking large amounts of alcohol.
Bacterial and viral infections cause problems such as mumps (a viral infection), in which the salivary glands are inflamed for about five days, and trench mouth (a bacterial infection). Tonsillitis (inflamed tonsils) can be caused by a virus or a bacterium.

LATE ADDITIONS 'Wisdom' teeth (in orange) grow through when a child becomes an adult, around the age of 18. Sideways growth can cause them to become impacted (righthand side).

Dentistry

A wide range of specialisations are included within dentistry, depending on the area and level of treatment:

Restorative dentistry repairs damaged teeth through processes such as drilling out decayed tooth material and replacing it with a filling (a silver-mercury amalgam, or more often a white synthetic resin), and fitting crowns (metal or porcelain 'caps') over teeth with worn surfaces.
Endodontics treats diseases of the root canal (the pulp-filled cavity in the root of a tooth).
Prosthodontics replaces lost teeth with bridges (artificial teeth attached to natural teeth on either side) and dentures.
Oral surgery involves operations on the mouth such as the removal of impacted wisdom teeth, implanting artificial teeth into the jaw and procedures such as repairing cleft palates.
Orthodontics corrects irregularities in the alignment and growth patterns of teeth.
Periodontics treats diseases of the jawbones and gums that support the teeth.

TEMPORARY TEETH Children have 20 'milk' teeth. As the 32 adult teeth grow, they push out the milk teeth.

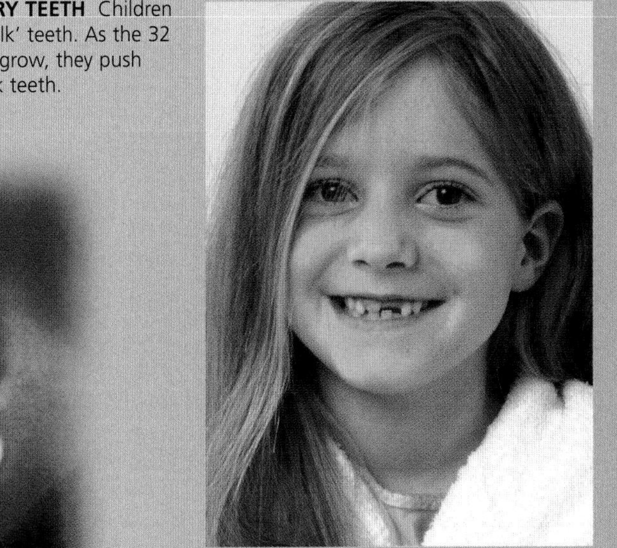

QUESTION NUMBER

The numbers or star following the answers refer to information boxes on the right.

ANSWERS

66 **A: Beans** (the others are complete proteins) ❹

138 **D: 40 tonnes** ❻

196 **False** (very small amounts are essential) ❸

222 **Protein deficiency** ❹

346 **Cholesterol** ❸

361 **Carbon, hydrogen and oxygen** ❷ ❸

362 **Starch** ❷

363 **Milk sugar and fruit sugar** ❷

398 **More** ❷ ❸

★ 473 **Water**

501 **Soup noodles** ❼

532 **Fibre** ❺

544 **Laxative** ❺

713 **C: Olive oil** ❸

727 **Nutrition** ❶

730 **Fibre** ❺

778 **True** ❺

805 **Roughage** ❶ ❺

806 **Vegetables** ❺

807 **Coeliac disease** ❽

824 **Amino acids** ❹

911 **Gluten** ❽

969 **False** ❺

Nutrition and diet 1

Core facts ❶

◆ The substances in food used to fuel the body are called nutrients; the scientific study of them is known as **nutrition**. The energy needed to fuel the body is measured in **kilocalories**.

◆ Carbohydrates, proteins and fats, known as **macronutrients**, supply energy for bodily processes and provide materials to build and renew body tissue.

◆ Vitamins and minerals are needed in much smaller quantities than macronutrients and are known as **micronutrients**. They are vital to processes such as bone growth and blood and fluid regulation.

◆ Dietary **fibre** (roughage) is not a nutrient, but helps to sustain healthy digestion.

◆ The body also needs **water** to function.

Carbohydrates ❷

The body's main source of energy is derived from starches and sugars – known as carbohydrates. One gram of carbohydrate yields about 4 kilocalories of energy. Carbohydrates do not provide any other nutritional benefits, though some carbohydrate-rich foods also contain dietary fibre.

◆ Carbohydrate molecules are made up of atoms of **carbon**, **hydrogen** and **oxygen**. Like water (H_2O), they contain twice as many hydrogen (H) atoms as oxygen (O) atoms – hence the 'hydrate' ('containing water') in carbohydrate, even though carbohydrates do not contain any actual water molecules.

◆ **Sugars** are called simple carbohydrates because they have small molecules. Glucose (blood sugar) is the main energy source for the body's cells, and can be absorbed from milk (in the form of lactose), sucrose (table sugar) and vegetables. Or it can be taken indirectly from fructose (fruit sugars), which is converted into glucose by the liver.

◆ **Starches** have large molecules built up from many linked sugar units, and are known as complex carbohydrates. Digestion breaks the molecules into their constituent units gradually, so the energy becomes available more slowly than that provided by sugars. Plants use starch molecules to store energy; dietary sources of starch include vegetables, fruits and grains.

STARCHY FOOD Cereal-based products such as pasta are high in carbohydrates.

Fats ❸

Fats also contain energy, but in a highly concentrated form. They provide the body's main energy store. One gram of fat yields 9 kilocalories. The body needs small amounts of certain fats, mainly to build cell membranes, but any extra is unnecessary and may be stored as adipose tissue.

◆ Fat molecules are made of **carbon**, **hydrogen** and **oxygen** atoms, in different proportions from carbohydrates. They comprise glycerol (glycerine) and chains of fatty acids, and digestion breaks them down into these constituent parts.

◆ Depending on the proportion of hydrogen atoms they contain, the fatty acid part is classified as **saturated** (found in butter, hard cheeses, meat products and fried foods), **monounsaturated** (found in avocados, olive and peanut oils, nuts and seeds) or **polyunsaturated** (found in fish oils, and in corn, safflower and sunflower oils).

◆ Eating a lot of fat can raise levels of **cholesterol** (a lipoprotein – a fat/protein combination) in the blood. There are two types of cholesterol: low-density lipoproteins (LDLs) and high-density lipoproteins (HDLs). LDLs form in response to the intake of saturated fats, and are associated with **heart disease**. HDLs, raised by the intake of monounsaturated fats, are thought to protect the arteries against disease.

◆ At least 25 g of fat a day, taken from polyunsaturated oils or full-fat dairy products, is necessary to supply the body with **vitamins** A, D, E and K. Fat also facilitates the absorption of these vitamins, which are known as **fat-soluble**.

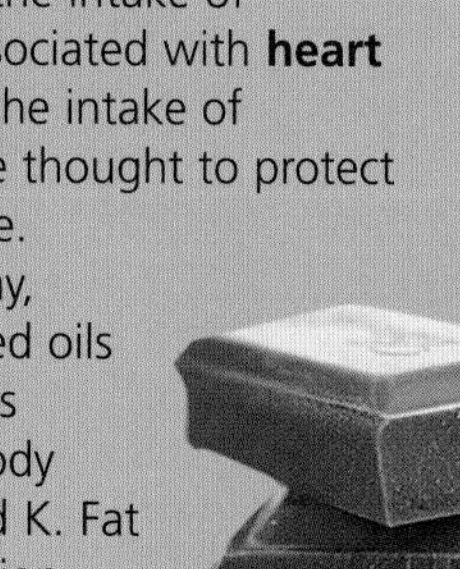

HIGH-ENERGY About 30 per cent of the weight of chocolate is fat, but it is ideal survival food for walkers and mountaineers.

Proteins

Proteins help to build the human body. For example, skin, muscle and cartilage are made mostly of protein; so are antibodies, many hormones, and the enzymes which regulate cellular activity. One gram of protein yields about 4 kilocalories of energy.

A HEALTHY SNACK Peanuts contain low-quality protein, but are rich in essential fatty acids.

◆ Protein molecules are complex, containing **carbon**, **hydrogen**, **oxygen** and **nitrogen** atoms, and sometimes iron, phosphorus and/or sulphur. They consist of one or more polypeptide chains. Each chain is made up of chemical components called **amino acids** – the end products of protein digestion.

◆ Twenty different amino acids are found in proteins. If necessary the body can create many amino acids from others in the diet, but eight (nine in young children) cannot be synthesised and must be obtained from food – these are known as **essential** amino acids.

◆ The **daily protein intake** recommended by the United Nations' food and health agencies is about 15 g (1/2 oz) for children under three, rising to 30-38 g (1-1 1/3 oz) for teenagers and adults. Pregnant women need an extra 6 g (1/4 oz) of protein a day.

◆ The best sources of protein – because they contain all of the essential amino acids – are **animal products** such as meat, fish, milk, cheese and eggs. **Plant foods** such as grains, pulses (peas, beans and lentils), nuts and many vegetables also contain protein, but these are incomplete – they lack some essential amino acids. Strict **vegetarians** should combine plant foods, such as grains and pulses, to obtain the right balance of essential nutrients.

◆ Protein is the only macronutrient whose lack can cause a deficiency disease. Inadequate protein leads to listlessness, poor disease-resistance, stunted growth and, in extreme cases, liver damage. An enlarged liver and fluid in the tissues causes the pot-belly typical of malnourished children in developing countries, a condition known as **kwashiorkor**.

Water

473

The human body is about 60 per cent water – the medium in which chemical reactions take place. Water is constantly lost through sweating, breathing and passing urine. If the water percentage drops by only a few points a person starts to feel listless. A drop to 47 per cent is fatal.

On average, an adult needs at least 2.5 litres (4 1/2 pt) of water a day, from food and drink. The kidneys help to maintain water balance: if water is short they produce small amounts of concentrated urine, retaining more fluid in the body.

Dietary fibre

Fibre, also known as roughage, is made up of **indigestible cellulose**, **pectin** and **gums** contained in the cell walls of plant foods. It adds bulk to the diet and helps to push waste through the digestive system, keeping it healthy. Fruits, vegetables and wholegrain foods are rich in fibre, but refined white bread also contains significant amounts.

WHOLE GRAIN Some fibres, such as brown rice, reduce the body's mineral absorption. Eating fibre from a range of foods avoids this.

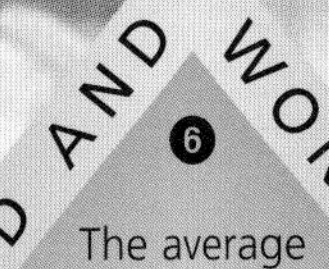

WEIRD AND WONDERFUL

The average **lifetime intake** for a person in the industrialised world is about 40 tonnes of food – equivalent to seven full-grown elephants – and about 65 000 litres (14 300 gallons) of water taken in food and drinks.

Food films

◆ **1986** *Tampopo* (set in noodle bar, Japan)
◆ **1987** *Babette's Feast* (village home, Scandinavia)
◆ **1989** *The Cook, the Thief, his Wife and her Lover* (restaurant, England)
◆ **1994** *Eat Drink Man Woman* (family homes, Taiwan)
◆ **1997** *Soul Food* (family homes, USA)
◆ **2000** *Chocolat* (chocolate shop, France)
◆ **2002** *Mostly Martha* (restaurant, Germany)

Diet diseases

Disease	Features
Coeliac disease	Gluten intolerance, leading to intestinal damage and an inability to absorb fat.
Coronary artery disease	Thickening of the artery walls associated with a high intake of saturated fat.
Marasmus	Wasting and weakness caused by malnutrition.
Osteomalacia	Soft bones caused by vitamin D deficiency (vitamin D helps calcium absorption).

QUESTION NUMBER

The numbers or star following the answers refer to information boxes on the right.

ANSWERS

72	**Limeys** ❸
171	**The Chinese** ❹
★ 221	**Iodine**
291	**Cod liver oil and orange juice** ❶
342	**Calcium** ❷
397	**Minerals and vitamins** ❶ ❷
427	**Potassium** ❷ ❺
533	**Iron** ❷
716	**A: Vitamins** ❶ ❹
873	**Carotene** ❶
893	**Vitamins** ❶
982	**The Boomtown Rats** – in UK charts for 11 weeks
985	**Orange Juice** – formed in 1978, disbanded 1985

Nutrition and diet 2

Vitamins

❶

Vitamins are vital to metabolic processes such as bone formation, blood-clotting and tissue repair. Vitamins A, D, E and K are fat-soluble, and can be stored in the body. Water-soluble vitamins – the B group and vitamin C – cannot be stored, and must be eaten regularly.

Vitamin	RDA*	Good sources	Function in body
A Retinol	0.8-1 mg	Egg yolk; fish oils; offal. Made by body from carotene in green, yellow and orange vegetables.	Growth; immunity; maintains health of mucous membranes; vision (especially in dim light).
B_1 Thiamine	1-1.4 mg	Beans; fortified cereals; fortified flour; nuts and seeds; pork; brown rice; wheat germ.	Carbohydrate metabolism; maintains health of muscular and nervous systems.
B_2 Riboflavin	1.2-1.6 mg	Avocados; fortified cereals; dairy products; eggs; nuts; offal; wheat germ; yeast.	Cellular respiration; tissue maintenance and repair.
B_3 Niacin	13-18 mg; excess harmful	Fish; fortified cereals (not natural maize) and flour; liver; meat; peanuts; yeast; wholegrains.	Carbohydrate metabolism; maintains health of skin, nerves and digestive system.
B_5 Pantothenic acid	4-7 mg	Fish; legumes; poultry; wholegrains; yoghurt. Also made by intestinal bacteria.	Carbohydrate, fat and protein metabolism; maintains health of nervous system.
B_6 Pyridoxine	2 mg	Bananas; fish; meat; potatoes. Also made by intestinal bacteria.	Protein metabolism; red blood cell and antibody formation.
B_9 Folic acid	0.2-0.4 mg	Meat; orange juice; pulses; green vegetables; wheat germ; wholemeal flour; yeast.	DNA production; cell division (with B_{12}) – including red and white blood cell formation.
B_{12} Cobalamin	0.003 mg	Fortified cereals; cheese; eggs; fish; meat; oysters.	DNA production; maintains health of nervous system; helps functioning of folic acid.
Biotin (B group; not numbered)	0.1-0.3 mg	Egg yolk; nuts; offal; soy products; wheat germ; yeast. Also made by intestinal bacteria.	Fat metabolism.
C Ascorbic acid	60 mg (100 mg for smokers)	Fruits and vegetables, especially raw blackcurrants, citrus fruit, tomatoes, peppers and potatoes.	Antioxidant (protects against disease); maintains health of connective tissue and cell walls.
D Calciferol	0.01 mg	Mainly sunlight on skin. Also full-fat dairy products; eggs; liver; fortified margarine; oily fish.	Facilitates absorption of calcium and phosphorus from intestines and bone growth.
E Tocopherol	8-10 mg	Butter; cereals; eggs; fortified margarine; nuts and seeds; wheat germ; vegetable oils.	Antioxidant (protects against disease). Thought to protect cell membranes.
K Menaquinone/ phylloquinone	0.07-0.14 mg	Leafy green vegetables. Also made by intestinal bacteria.	Facilitates blood-clotting.

*Recommended daily amount for adults.

★ 221

Iodine

Small amounts of iodine are vital for making thyroid hormones, and a lack of iodine causes **thyroid overactivity**. In places where the diet is lacking in iodine – because there is little seafood or the soil (and thus food and water) contains no iodine – people can develop a swollen thyroid gland, known as a goitre.

CONVENIENCE FOOD
Bananas are a good source of potassium. They are also rich in starch, making them an ideal energy snack.

LIFE SAVER Oily fish contain omega-3 fatty acids, which protect against heart disease and blood clots.

Minerals

Minerals are inorganic nutrients used in processes such as bone growth and the generation of nerve impulses. Some, known as trace elements or microminerals, are needed in only very small amounts (much less than a gram per day), and can be poisonous in excess.

Mineral	RDA*	Good sources	Function in body
Calcium	0.8-1.2 g	Dairy products; dark green leafy vegetables.	Blood clotting; maintenance of bones and teeth; nerve and muscle function.
Chromium †	0.05-0.2 mg	Cereals; meat; offal; wholegrain flour; yeast.	Carbohydrate metabolism.
Cobalt †	Not specified	Eggs; meat; offal.	Red blood cell formation.
Copper †	2-3 mg	Fish; liver; meat; nuts; pulses; shellfish; wholemeal bread.	Enzyme function; red blood cell formation.
Fluorine †	1.5-4 mg	Small fish, eaten whole; tea; toothpaste; fluoridated water.	Hardening of tooth enamel (excess may discolour teeth).
Iodine †	0.1-0.15 mg	Iodised salt; seafoods; foods grown on soil containing iodine.	Manufacture of thyroid hormones.
Iron †	10-20 mg	Fortified cereals and flour; eggs; dried fruits; red meat; nuts; offal; pulses; shellfish.	Essential part of haemoglobin (oxygen-carrying molecules) in red blood cells.
Magnesium	0.25-0.7 g	Nuts and seeds; pulses; green leafy vegetables.	Bone and tooth formation; enzyme function; nerve and muscle function.
Manganese †	2-3 mg	Dried fruit; pulses; wholegrains.	Enzyme function.
Molybdenum †	0.15-0.5 mg	Barley; liver; oats; pulses.	Enzyme function.
Phosphorus	0.8-1.5 g	Wholegrain cereals; dairy products; fish; meat; nuts; yeast.	Bone and tooth formation; fat metabolism.
Potassium	2.5 g	Bananas; meat; milk; oranges; potatoes; yoghurt.	Carbohydrate metabolism; nerve and muscle function.
Selenium †	0.05-0.2 mg	Wholegrain cereals and flour; eggs; offal; seafood; yeast.	Antioxidant (protects against disease); thyroid function.
Sodium	2.5 g	Salt; most other foods.	Nerve and muscle function.
Sulphur †	Not specified	Eggs; milk; meat; pulses and all other protein-rich foods.	Manufacture of proteins, including insulin.
Zinc †	15 mg	Cereals; wholegrain flour; meat; nuts; pulses; shellfish; yeast.	Enzyme function; growth; immunity; protein metabolism; tissue healing.

*Recommended daily amount for adults (excess can be harmful). † Trace elements.

TIMESCALE

▶ **c.400 BC** Hippocrates suggests that liver (containing vitamin A) can cure blindness.
▶ **AD c.1100** Chinese sailors use citrus fruits to combat scurvy.
▶ **1754** Scottish naval surgeon James Lind recommends the use of citrus juice to treat scurvy.
▶ **1882** Japanese doctor Kanehiro Takaki adds meat and vegetables to a rice diet to cure beriberi (B vitamin deficiency).
▶ **c.1900** Dutch scientist Christiaan Eijkman proves that eating unpolished rice cures beriberi.
▶ **c.1906** British scientist Frederick Hopkins proposes a theory that foods contain 'accessory' factors (later known as vitamins) that are vital to health.
▶ **1912** Polish biochemist Casimir Funk proposes that a lack of 'vitamines' ('vital amines' – necessary nutrients) causes disease.
▶ **1918** British physician Edward Mellanby cures rickets with cod-liver oil.
▶ **1928** Hungarian researcher Albert Szent-Gyorgyi isolates vitamin C.
▶ **1948** Discovery of final proven vitamin (B_{12}) by British and US scientists.

WEIRD AND WONDERFUL

The Royal Navy began using **lemon-juice** rations to counteract scurvy (vitamin C deficiency) in 1795. They later switched to lime juice – hence the nickname 'limeys' – though limes contain much less vitamin C than lemons.

COMBINING FOODS Eggs contain iron, but in a form that the body cannot absorb easily. Eating foods rich in vitamin C with eggs helps iron absorption.

Sodium and potassium

Two of the most common food minerals, potassium and sodium, are vital to the functioning of the body. Body cells contain a relatively high concentration of **ions** (charged atoms) of potassium. Most sodium ions are outside the cells, in the blood and intercellular fluid.

The amounts of potassium and sodium are kept in careful balance. If too much potassium gathers outside the cells, **nerves** cannot transmit signals and the **heart** may beat irregularly or stop. Imbalances are regulated by the **kidneys**. For example, if sodium is lost through sweating, they excrete potassium and conserve sodium.

Dietary intake of sodium aids fluid retention, which increases **blood** volume and pressure. A high intake can contribute to chronic hypertension (high blood pressure).

QUESTION NUMBER

The numbers or star following the answers refer to information boxes on the right.

ANSWERS

2	**Anorexia (nervosa)** 6
73	**Monosodium glutamate** 3
103	**True** 1
★ 252	**B: William Hay** (others are high-fat, low-carbohydrate)
276	**Colourings** 3
474	**Europe** 3
514	**Metabolism** 2
622	**Anabolism** 2
695	**Jean Harlow** – directed by George Cukor
745	***Oliver Twist*** – first published in 1837
831	**B: 4000 kcal** 1
874	**Marine algae, or seaweed** 3

Nutrition and diet 3

Measuring energy 1

The energy content of food is measured in kilojoules (kJ) or **kilocalories** (kcal), also known as **Calories**. The term is often incorrectly shortened to 'calories' – there are 1000 'small' calories in a kilocalorie or large Calorie. One kcal is equivalent to 4.2 kJ.

Energy needs depend on body size and activity. A small woman working at a desk may 'burn' (use up) as few as 1500-1800 kcal (6300-7500 kJ) a day, whereas a large-framed manual workman may burn three times as much. In comparison, a 100-watt light bulb consumes 2060 kcal (8650 kJ) in 24 hours.

Activity	Energy used per hour
Badminton	340 kcal (1428 kJ)
Climbing stairs	620 kcal (2604 kJ)
Cycling (fast)	660 kcal (2772 kJ)
Football	540 kcal (2268 kJ)
Gardening (heavy)	420 kcal (1764 kJ)
Gardening (light)	270 kcal (1134 kJ)
Golf	260 kcal (1092 kJ)
Housework	270 kcal (1134 kJ)
Lying in bed	60 kcal (252 kJ)

Activity	Energy used per hour
Mountaineering	640 kcal (2688 kJ)
Squash	600 kcal (2520 kJ)
Standing still	120 kcal (504 kJ)
Swimming (fast)	720 kcal (3024 kJ)
Tennis	480 kcal (2016 kJ)
Walking (brisk)	300 kcal (1260 kJ)

BURNING CALORIES Jogging uses up 630 kcal (2646 kJ) of energy per hour – more than playing tennis or squash.

Metabolism 2

Billions of chemical reactions, known collectively as the metabolism, sustain the human body. Metabolic processes work in two ways:

- **Anabolic metabolism** builds simple molecules into complex molecules such as proteins using energy from food.
- **Catabolic metabolism** breaks down large molecules into smaller, simpler ones, at the same time releasing energy.

The rate at which the metabolism 'ticks over' when resting – the **basal metabolic rate (BMR)** – varies depending on a person's build and usual level of physical activity. People with a high BMR burn up energy faster, whatever exercise they take.

Food additives 3

Processed food contains additives such as preservatives and sweeteners. Some are synthetic; others are refined natural substances. EU countries must only use approved additives with an E (European Union) number, and list them on the label.

E number	Purpose	Example
E100s	Colourings	E102 tartrazine
E200s	Preservatives	E210 benzoic acid
E300s	Antioxidants and acidifiers	E338 phosphoric acid
E400s	Emulsifiers, stabilisers and thickeners	E406 agar (made from marine algae)
E500s	Acidity regulators and anticaking agents	E551 silica
E620-640	Flavour enhancers	E631 sodium inosinate
E900-949	Non-specific	E901 beeswax
E950-970	Sweeteners	E954 saccharin

★ 252

Diet gurus

Unconventional slimming methods include **high-fat, low-carbohydrate diets** such as the Atkins diet, which claims to 'burn' fat by restricting carbohydrates. Critics claim that this depletes nutrients and may cause heart disease. **Food-combining diets** – for example the Hay diet – claim that proteins and carbohydrates are digested differently, so should not be eaten together. Research suggests diets only work by reducing calorie intake.

What's on your plate? ❹

Different foods contain different balances of nutrients, and a balanced diet that includes a mixture of all food components should help to ensure long-term health. To achieve what constitutes a 'balanced diet', nutritionists have identified several main food groups and suggested an ideal daily amount of each group.

FRUIT AND VEGETABLES provide vitamins and minerals, carbohydrates and dietary fibre. Some, such as pulses and root vegetables, also provide protein. Five portions a day are recommended.

MEAT, POULTRY, FISH, EGGS, NUTS AND PULSES are good sources of protein, iron and some vitamins, but animal foods also contain substantial amounts of saturated fats. Three fairly small portions a day are recommended.

BREAD, PASTA AND CEREALS provide a good source of complex carbohydrates, some protein and (especially if wholegrain) dietary fibre. Nutritionists recommend four substantial helpings a day.

DAIRY PRODUCTS such as milk and cheese provide protein, calcium and some vitamins, but – unless very low in fat – they also contain large amounts of saturated fats. One to two low-fat portions a day are usually recommended. Some nutritionists include eggs in this group.

FATS, OILS AND SWEETS provide essential fatty acids (in fats and oils) and energy, but little else. They are needed only in very small quantities. Oily fish and olive oil are better sources of essential fats than butter, margarine or other spreads.

Overeating ❺

When calorie intake exceeds energy output, fat is stored in the body's cells. This can lead to **obesity**, when the body is more than 30 per cent heavier than its ideal weight. For women, the ideal balance is about 25 per cent of body weight stored as fat; for men, it is 15 per cent. The only remedy for obesity is to restrict calorie intake and gently increase energy expenditure.

Binge eating, in which 3000 to 6000 calories are consumed at one sitting, is a feature of the psychiatric condition **bulimia nervosa**. Sufferers starve themselves, binge, then induce vomiting. Persistent vomiting depletes potassium – treatment of bulimia includes a potassium-rich diet and counselling.

Undereating ❻

In countries where food is plentiful, undereating is often the result of isolation in old age or conditions such as **alcoholism** or **anorexia nervosa**. Low food intake causes nutritional deficiencies which exacerbate the problem – for example, lack of zinc depresses the appetite and dulls the sense of taste.

An alcoholic obtains a supply of calories from drink. This maintains energy levels but leads to nutritional deficiencies – particularly of B_{12} and thiamine, which can damage the nervous system. Sufferers of anorexia obsessively restrict food intake in an attempt to achieve a perceived ideal of extreme slimness. Treatment involves careful nutrition and counselling.

QUESTION NUMBER

The numbers or star following the answers refer to information boxes on the right.

ANSWERS

43	**9 m (30 ft)** ❷
97	**Alimentary canal** ❶ ❷
★ 260	**B: The stomach wall**
268	**The stomach** ❷
309	**Part of the small intestine** ❷
310	**Gullet** ❷
364	**Enzymes** ❶ ❷ ❻
365	**Sugars** ❶
366	**Fatty acids and glycerol** ❶
498	**False** ❹
502	**Gastric juice and bile** ❷ ❻
515	**Small intestine** ❷
516	**Large intestine** ❷
526	**C: A ring of muscle** ❷
534	**The pancreas** ❸
675	**Tiny finger-like projections** ❷
687	**False** – it is a term for nausea
721	**The stomach** ❷
731	**Gastroenteritis** – caused by food, drugs, allergies
732	**Stomach ulcer** ❹
764	**Lining of the stomach** ❷
935	**True** ❺
942	***Heartburn*** – released in 1986

Digestive system

Core facts ❶

◆ Most food cannot be used by the body in the form in which it is eaten. During the process of digestion, **enzymes** break it down into its simple chemical components

◆ **Carbohydrates** are digested into simple sugars, **proteins** into amino acids, and **fats** into fatty acids and glycerol.

◆ The process of digestion is carried out by the **alimentary canal** and its associated glands and organs. The digestive system is also responsible for **removing waste**.

◆ The alimentary canal is a long tube leading from the mouth to the anus, incorporating the oesophagus, stomach and intestines.

Alimentary canal ❷

The alimentary canal of an adult measures up to 9 m (30 ft) long. It has a muscular wall which pushes food along by a squeezing wave action called **peristalsis**.

The complete canal is lined by mucous membrane. Much of it is lined with finger-like projections called **villi**, 0.5-1.5 mm (1/50-1/18 in) long, which are covered in even smaller projections called **microvilli**. This increases the surface area available for nutrient absorption.

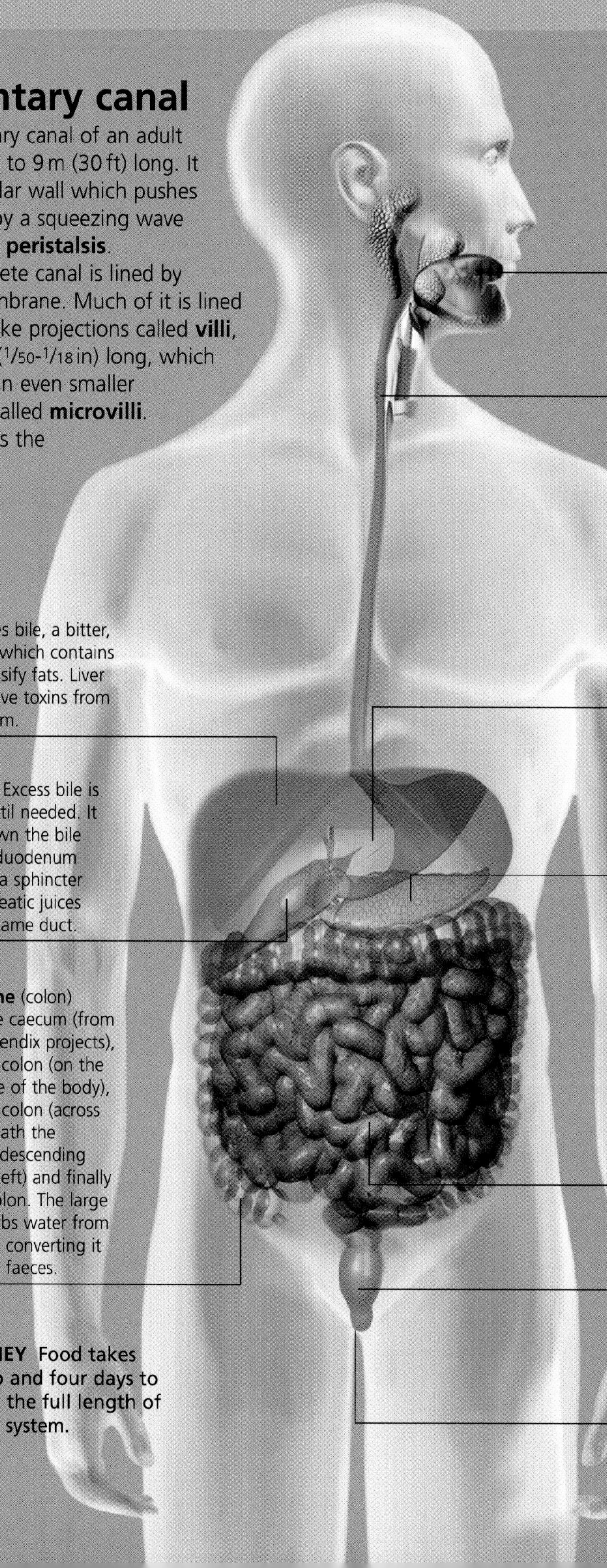

Mouth Teeth cut and grind food and mix it with saliva. Enzymes in saliva begin digestion. The tongue shapes food into a bolus and pushes it towards the throat.

Oesophagus (gullet) Peristalsis propels the bolus (soft, moist ball) of food towards the stomach.

Stomach Glands in the stomach walls produce a gastric juice of hydrochloric acid and digestive enzymes to start the digestion of proteins. The muscular walls churn the contents into a paste called chyme. The pyloric sphincter muscle controls the flow of chyme into the small intestine, while preventing backflow.

Liver Produces bile, a bitter, alkaline liquid which contains salts that emulsify fats. Liver cells also remove toxins from the bloodstream.

Gall bladder Excess bile is stored here until needed. It then flows down the bile duct into the duodenum (controlled by a sphincter muscle). Pancreatic juices flow into the same duct.

Pancreas Produces insulin, which regulates blood sugar. Pancreatic juice contains digestive enzymes and sodium bicarbonate, which neutralises stomach acid.

Small intestine A three part tube comprising the duodenum, jejunum and ileum. Chyme moves through it by peristalsis. Enzymes in the intestinal wall complete the digestion of proteins, carbohydrates and fats. Digested molecules are absorbed into the blood and fatty molecules are absorbed into the lymph.

Large intestine (colon) Consists of the caecum (from which the appendix projects), the ascending colon (on the right-hand side of the body), the transverse colon (across the body beneath the stomach), the descending colon (on the left) and finally the sigmoid colon. The large intestine absorbs water from digested food, converting it into semi-solid faeces.

Rectum Holds waste material until it is expelled through the anus.

Anus A sphincter muscle controls the expulsion of waste material (faeces).

LONG JOURNEY Food takes between two and four days to pass through the full length of the digestive system.

Pancreas

3

The pancreas plays a vital role in **digestion** and in balancing **blood sugar** levels. Gastric secretions and chyme (see box 2) in the small intestine stimulate grape-like clusters of pancreatic cells, known as **acini**, to produce pancreatic juice – a mixture of digestive enzymes which flows through the bile duct into the intestine.

The pancreas also contains about 1 million **islets of Langerhans** – groups of endocrine cells which secrete insulin, a hormone that stimulates glucose uptake from the blood.

ENZYME FACTORY Clusters of acini (in blue) are connected by small ducts which join up with a central pancreatic duct.

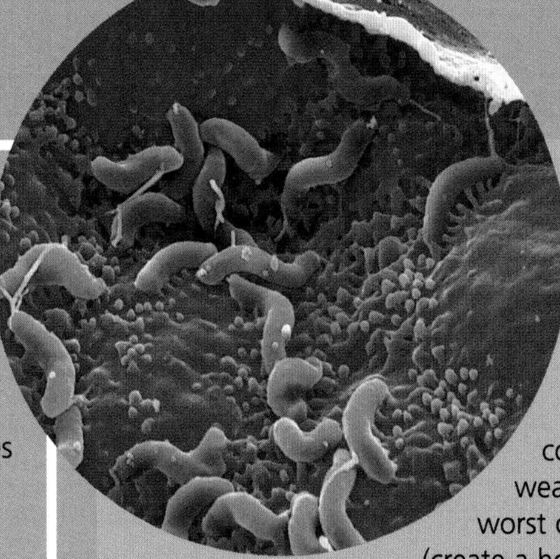

IDEAL HOME The bacteria *Helicobacter pylori* thrive in an acid environment.

4

The ulcer revolution

Peptic ulcers occur when the mucus coating of the stomach or duodenum wears away, creating a 'raw' patch. In the worst cases, an ulcer can bleed or perforate (create a hole), causing anaemia and vomiting. Until the 1980s, ulcers were thought to be caused by the over-production of **gastric acid**, and treatment aimed to neutralise the acid. It has since been dicovered that ulcers often result from the action of a bacterium – ***Helicobacter pylori***. These bacteria are found in the stomach of almost 50 per cent of people, whether or not they have an ulcer, and in virtually everyone who does have an ulcer. The same bacteria increase the risk of developing **stomach cancer**. It is now possible to cure many (although not all) ulcers and reduce the rate that stomach cancer spreads with antibiotic treatment.

X-RAY EVIDENCE A persistent peptic ulcer (top right, in black) may be treated with surgery.

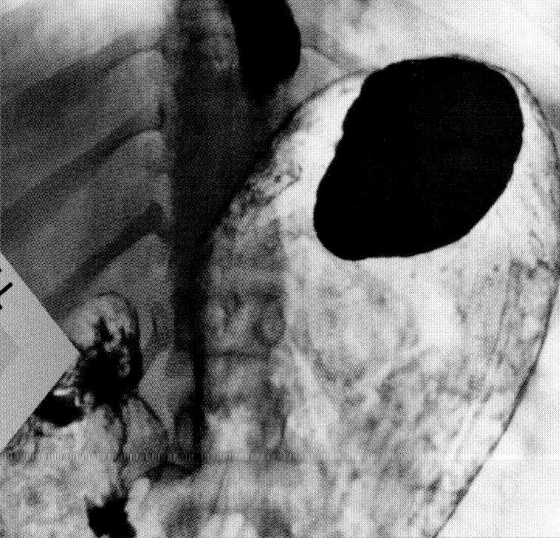

WEIRD AND WONDERFUL

5

The alimentary canal is mostly protein, so why does it not digest itself? Glands in the walls of the canal produce large amounts of **thick mucus**, forming a coating which protects it against acids and digestive enzymes.

Digestive enzymes

6

The chemical catalysts that drive digestion are the enzymes produced in the mouth, stomach, pancreas and intestines.

◆ **Saliva** contains **amylase**, which digests starch and glycogen into maltose.

◆ The **gastric juices** of the stomach contain **pepsin**, which digests proteins into smaller units called peptones.

◆ **Pancreatic juice** contains **amylase**; **lipase**, which digests fat droplets into fatty acids and glycerol; and **trypsinogen** and **chymotrypsinogen** – inactive enzymes which become activated in the intestine (this prevents them digesting the tissue which produces them).

◆ In the intestine, **enterokinase** converts trypsinogen into its active form, trypsin, which digests peptones into smaller units known as peptides. Trypsin also converts chymotrypsinogen into **chymotrypsin**, which digests some peptides into amino acids. **Lactase**, **maltase** and **sucrase** digest the sugars lactose, maltose and sucrose into substances such as glucose, **lipase** digests fats and **peptidase** digests peptides into amino acids.

★ 260

Beaumont's window

In 1822, a patient of US army surgeon **William Beaumont** took a musket wound in the stomach. He recovered, but the wound stayed open and had to be plugged. Beaumont was able to study **gastric activity** through the opening, known as 'Beaumont's window', making the earliest detailed investigation of the stomach.

Digestive disorders

7

Inflammatory bowel diseases (IBDs), unlike infections, often have no obvious cause.

Disorder	Features
Appendicitis	Inflammation of the appendix. Causes pain and vomiting.
Cancer	Intestinal tumours. Can arise from genetic tendency.
Food poisoning	Infection from contaminated food. Causes diarrhhoea and/or vomiting.
Hernia	Part of the intestine protrudes through the abdominal wall.
Crohn's disease	Thickened/ulcerated intestines. Causes pain, diarrhoea, fever and weight loss.
Ulcerative colitis	Inflamed large intestine. Causes pain, diarrhoea, fatigue, fever, weight loss and haemorrhages.

QUESTION NUMBER — The numbers or star following the answers refer to information boxes on the right.

ANSWERS

16	**Jaundice** ❺
91	**The liver** (choler = bile) ❶❷
112	***King Lear*** – lily-livered means to lack courage.
261	**The kidneys** ❶❹
262	**The bladder** ❶❹
316	***Blackadder*** – series III, episode 3
341	**Cystitis** ❺
367	**Detoxification** ❶❷
368	**Urea** ❹
413	**Human liver** ❶❷
420	**The Tintin books** – created by Hergé in 1929
503	**Into the bloodstream** ❶❷❸
504	**(a) Ureter, (b) Urethra** ❶❹
535	**Excretion** ❶❹
536	**The kidneys** ❶❹
537	**The liver** ❶❷
623	**Ammonia** ❷
657	**Hepatitis** ❺
676	**The liver** ❶❷
677	**The kidneys** (from Greek *Nephros*, for kidney) ❹❺
733	**Kidney stone** ❺
925	**Cirrhosis** ❺
★ 926	**Dialysis**

Liver and kidneys

Core facts ❶

◆ The liver and kidneys are organs of **excretion**: they eliminate waste from the body.

◆ The liver (the **hepatic system**) excretes waste in bile, which enters the small intestine and helps to digest fats. The kidneys, ureters, bladder and urethra (the **renal system**) excrete waste dissolved in water in the form of **urine**.

◆ The liver is the body's main chemical processing organ. It **detoxifies** harmful substances, manufactures **blood proteins** and recycles material from dead red blood cells.

◆ The kidneys regulate the body's **fluid levels** by filtering the blood. They also regulate the production of **red blood cells**.

A chemical factory ❷

The liver is one of the largest and most complex organs in the body, weighing on average 1.5 kg (more than 3 lb) in an adult. Its cells, **hepatocytes**, are involved in hundreds of metabolic tasks, mainly involving digestion and blood regulation.

The liver produces 600-1000 ml (1-1¾ pints) of fat-digesting **bile** a day; it converts digested sugars into **glucose**; and it stores glucose and vitamins. It also **detoxifies** chemicals and converts ammonia, a poisonous waste metabolic product, into urea for excretion.

The blood-clotting protein **fibrinogen** is made in the liver. The hepatocytes also break down **haemoglobin** from dead red blood cells, store the iron for re-use and excrete the waste in bile.

Vena cava Returns blood to the heart.

Hepatic vein Returns blood to the vena cava.

Hepatic duct Drains bile from liver.

Cystic duct Connects gall bladder to bile duct.

Gall bladder Temporarily stores bile.

Hepatic portal vein Brings blood from intestines.

Hepatic artery Brings oxygen-rich blood from aorta.

Common bile duct Leads to duodenum (part of small intestine).

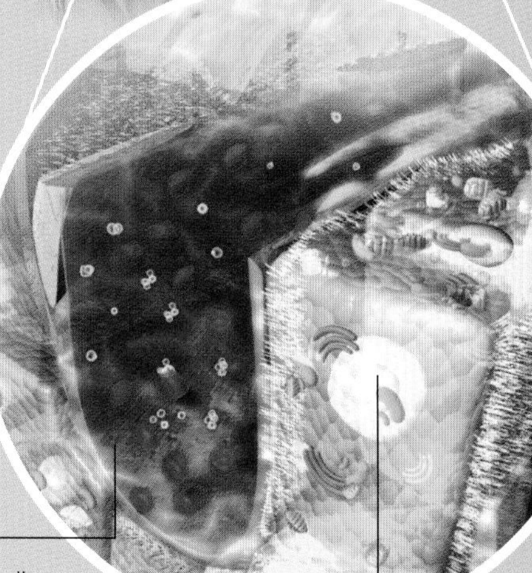

IN DETAIL Liver tissue is composed of up to 100 000 lobules, about 1 mm (1/25 in) across, full of blood-filled cavities called sinusoids (right).

Sinusoids Cavities fed with both arterial and portal vein blood.

Hepatocytes Specialised cells which carry out metabolic processes.

Gall bladder ❸

The **gall bladder** acts as a storage facility for 40-70 ml (1½-2½ fl oz) of bile from the liver. Its mucous lining absorbs water and electrolytes (electrically charged particles) from the bile, making it 5-10 times more concentrated.

The gall bladder is stimulated to contract by the hormone **cholecystokinin**, produced in the upper intestine. Bile then flows into the duodenum, where it helps to digest fats.

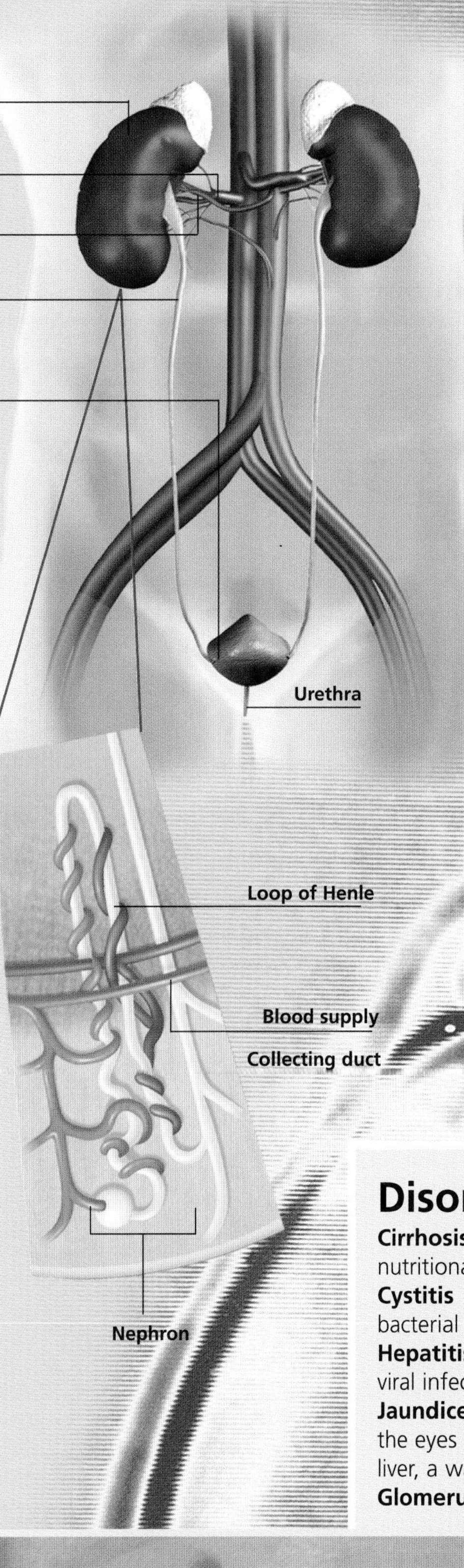

Kidney Filters waste products.

Renal vein Drains blood from kidney.

Renal artery Feeds blood to kidney.

Ureter Drains urine from kidney.

Bladder Stores urine until it is expelled through the urethra (the tube leading out of the body from the bladder).

Urinary system ❹

The kidneys lie on each side of the small of the back, and are the body's main excretory organs. Inside the kidneys, a two-stage filtering process produces **urine** – a mixture of **urea**, **uric acid** and other waste.

Filtering takes place in **nephrons**. These are tiny tubules beginning with a globular structure (known as a glomerular capsule) enclosing a knot of capillaries. Blood pressure forces plasma out through the capillary walls into the nephron. Here useful substances such as **amino acids** and **glucose** are reabsorbed into the blood, along with as much **salt** and **water** as the body needs. The remaining fluid (urine) filters through collecting ducts into the ureters and the bladder.

The kidneys also secrete erythropoietin, an enzyme which regulates **red blood cell production** in the bone marrow.

FILTERING TUBE Each kidney contains about 1.3 million nephrons (right). The blood vessels absorb and recirculate nutrients and water.

★ 926

Dialysis

If the kidneys stop working, artificial filtering, or dialysis, has to be used to detoxify the blood. In **haemodialysis**, blood from a vein passes into the dialysis machine and is filtered through a membrane. The cleansed blood is fed back into the body. **Peritoneal dialysis** uses the peritoneum – the lining of the abdomen – as a filter. Dialysis fluid is fed through a catheter into the abdominal cavity. Waste passes into the fluid, which is then drained through the catheter and replaced.

Disorders and diseases ❺

Cirrhosis Irreversible liver scarring caused by nutritional deficiency, alcohol abuse or infection.
Cystitis Inflammation of the bladder, caused by bacterial infection.
Hepatitis Inflammation of the liver caused by a viral infection or alcohol abuse.
Jaundice Yellowing of the skin and the whites of the eyes caused by a build-up of bilirubin in the liver, a waste product of haemoglobin break down.
Glomerulonephritis Inflammation of the glomerular capillaries in the kidneys through irritation by foreign material.
Kidney failure The kidneys stop working properly – suddenly, after severe shock, or gradually, as a result of conditions such as diabetes.

HARD PROBLEM Deposits in urine such as calcium can form stones in the kidneys (shown in orange), causing intense pain.

Brain and nervous system 1

QUESTION NUMBER

The numbers or star following the answers refer to information boxes on the right.

ANSWERS

- **61** C: Meninges ❷
- ★ **104** True
- **139** C: 2400 cm^2 (*c.*2½ sq ft) ❷
- **244** The cerebellum ❷
- **406** D: 100 billion ★
- **407** C: 100 000 ❹
- **505** Ventricles ❷
- **517** Grey matter ❺
- **525** A: A neurotransmitter ❻❼❽
- **529** A: A synapse ❻
- **556** Central nervous system ❶❷
- **678** Cortex ❷
- **688** False ('white matter' has myelin) ❺
- **770** Brain folds ❷
- **777** True (they connect via the corpus callosum) ❸
- **837** D: 20 per cent ❷
- **851** True ❸
- **894** Spinal cord ❶❷
- **932** True (that's 180 km/112 miles per hour) ❽
- **953** B: 0.1 volts ❽

Core facts ❶

◆ The **central nervous system** (CNS) consists of the brain and the spinal cord running down the centre of the vertebrae. It controls conscious and unconscious body functions.

◆ The nerves that extend from the spinal cord throughout the body make up the **peripheral nervous system** (PNS), which transmits signals between the body and the brain and spinal cord.

◆ Nervous tissue is composed of bundles of single-celled nerve fibres. **Electrical signals** transmit messages along the fibres to and from the brain and the body's muscles and organs.

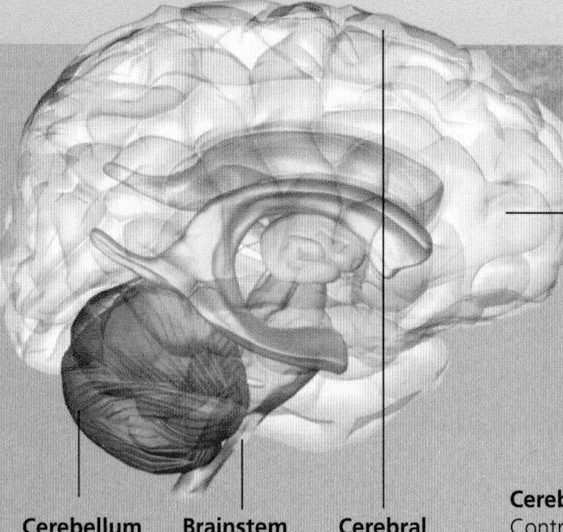

Central nervous system ❷

The brain weighs just 1.25-1.4 kg (2¾-3 lb) but uses 20 per cent of the body's oxygen supply. Its workings are not fully understood, but certain areas are known to control certain activities. It is connected to the rest of the body by the spinal cord.

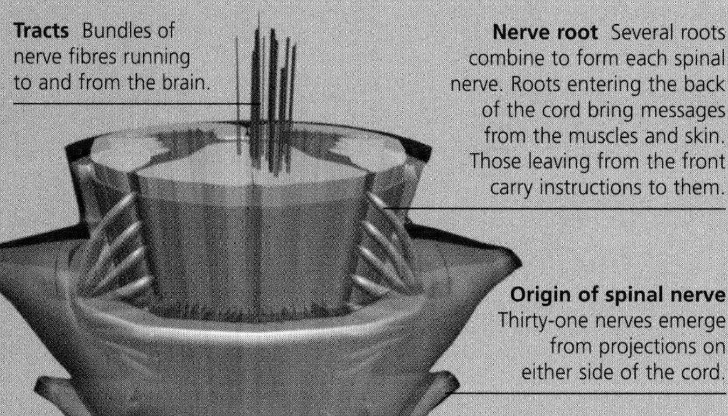

CONTROL CENTRE
The deeply folded outer surface of the adult brain (above) would measure about 2400 cm^2 (2½ sq ft) if opened out flat. The spinal cord leading from the brainstem consists of bundles of long nerve fibres, or axons, which connect the brain with the rest of the body.

Left vs right

3

Most nerves feeding signals from body to brain cross over in the brain stem – the left cerebral hemisphere is linked to the right-hand side of the body, and vice-versa. The exceptions are some of the optic nerves, and all of the olfactory nerves from the nose.

The left hemisphere of the brain is dominant in **speaking**, **writing** and **analytical skills**. The right hemisphere is the main centre for **creative pursuits** such as art and music.

4

LOST CONNECTIONS Human brain tissue reaches its maximum mass around the age of 20, then decreases. Up to 100 000 brain cells (background picture) die every day.

Structure of a neuron

6

Neurons (nerve cells) have three sections:

- a **cell body** about 0.025 mm (1/1000 in) across;
- branching **dendrites** about 0.5 mm (1/50 in) long that receive nerve signals;
- an **axon** (nerve fibre) which transmits signals. Axons measure from less than 1 mm (1/25 in) to 1 m (3 1/4 ft) long.

IN RELAY Electrical signals are passed from cell to cell by chemical substances called neurotransmitters. The chemicals are released across a gap, known as a synapse, between one cell and the next.

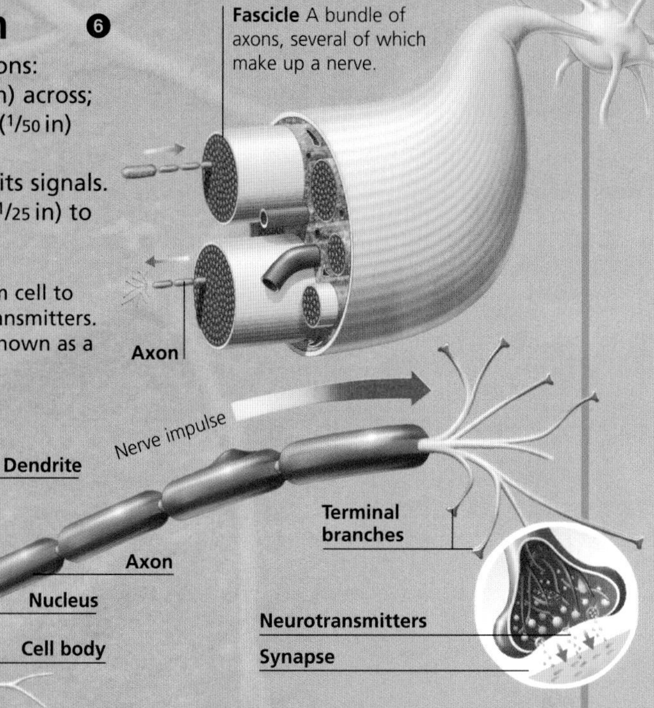

Tie-breaker

7

Q What is the name of the neurotransmitter that reduces anxiety and elevates mood?

A Serotonin. Low serotonin levels can cause depression. Many anti-depressant drugs encourage higher levels of serotonin to accumulate in the synapses. Excessive serotonin can cause migraine and nausea.

★ 104

Nerve connections

★

One neuron can link up with as many as 1000 others. A single brain neuron has up to 50 000 branches. Allowing for duplicate connections, it could communicate with 250 000 other nerve cells. There are some 100 billion neurons in the brain, making a huge number of signalling pathways.

Nerve signals

8

At rest, nerve cells have a negative electrical charge, and the extracellular fluid surrounding them has a positive charge. A nerve signal begins when neurotransmitters or sensory receptors stimulate the thin cell membrane, causing a change in its permeability and allowing **sodium ions** (electrically charged atoms) to pass through it into the cell and **potassium ions** to pass out of the cell.

The movement of ions creates an electrical disturbance which travels along the nerve fibre (axon) at an **average speed** of about 50 m per second (165 ft/sec) – signals travel at more than twice this speed in the fastest nerves, and at less than 1 m per second (3 ft/sec) in the smallest fibres, which do not have an insulating myelin sheath.

The electrical disturbance (or impulse) is known as an **action potential**, and nerve signals consist of a series of such impulses. A single nerve fibre can 'fire' between 50 and 300 times a second, returning to its resting state after each impulse. An impulse creates a voltage of up to 100 millivolts (1/10 volt), and always travels from dendrite to axon.

Grey and white matter

5

Areas in the cerebral cortex and the core of the spinal cord contain a high density of cell bodies and dendrites, which makes them appear grey. This tissue is known as **grey matter**. In deep areas of the cerebral cortex dense with axons, the tissue is known as **white matter** – axons are sheathed in myelin, a fatty substance which looks white.

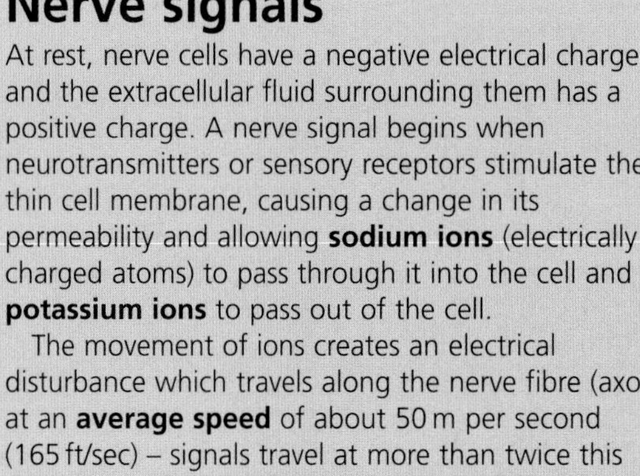

DENSE FIBRES A cell body (coloured blue) in a mass of dendrites forms grey matter.

Brain and nervous system 2

QUESTION NUMBER

The numbers or star following the answers refer to information boxes on the right.

ANSWERS

15 **Epilepsy** 7

26 **True** 1 5

44 **12 pairs** 1

45 **31 pairs** 1

560 **Transient ischaemic attack** 3 7

615 **Reflex action** 5

715 **B: Cranial nerves** 1

734 **Alzheimer's** 7

★ 754 **Autonomic nervous system**

889 **Andreas Vesalius** 6

912 **The brain** – usually caused by a viral infection

930 **Meningitis** 7

951 **C: Motor neurons** 2 5

952 **C: The face** 7

983 **Duran Duran** – in 1984

984 **Elvis Presley** (1935-77)

996 **Epilepsy** 7

Peripheral nervous system

1

Nerves extend in pairs from either side of the central nervous system to all parts of the body. Twelve pairs of **cranial nerves** connect with the underside of the brain. These include:

◆ **Optic and olfactory nerves** carrying visual and scent signals.
◆ Nerves carrying sensations and control signals to and from other parts of the head.
◆ The **vagus nerve**, which relays signals to control processes such as breathing, heart rate and speech.

From the spinal cord, 31 pairs of nerves emerge through gaps between the vertebrae. They are grouped according to the section of the spine and parts of the body they connect with:

◆ **Cervical nerves** connect with the neck, shoulders, arms and hands.
◆ **Thoracic nerves** connect with the chest, back and upper abdomen.
◆ **Lumbar nerves** connect with the lower back and parts of the legs.
◆ **Sacral nerves** connect with the genitals, buttocks, legs and feet.
◆ **Coccygeal nerves** connect with the base of the back.

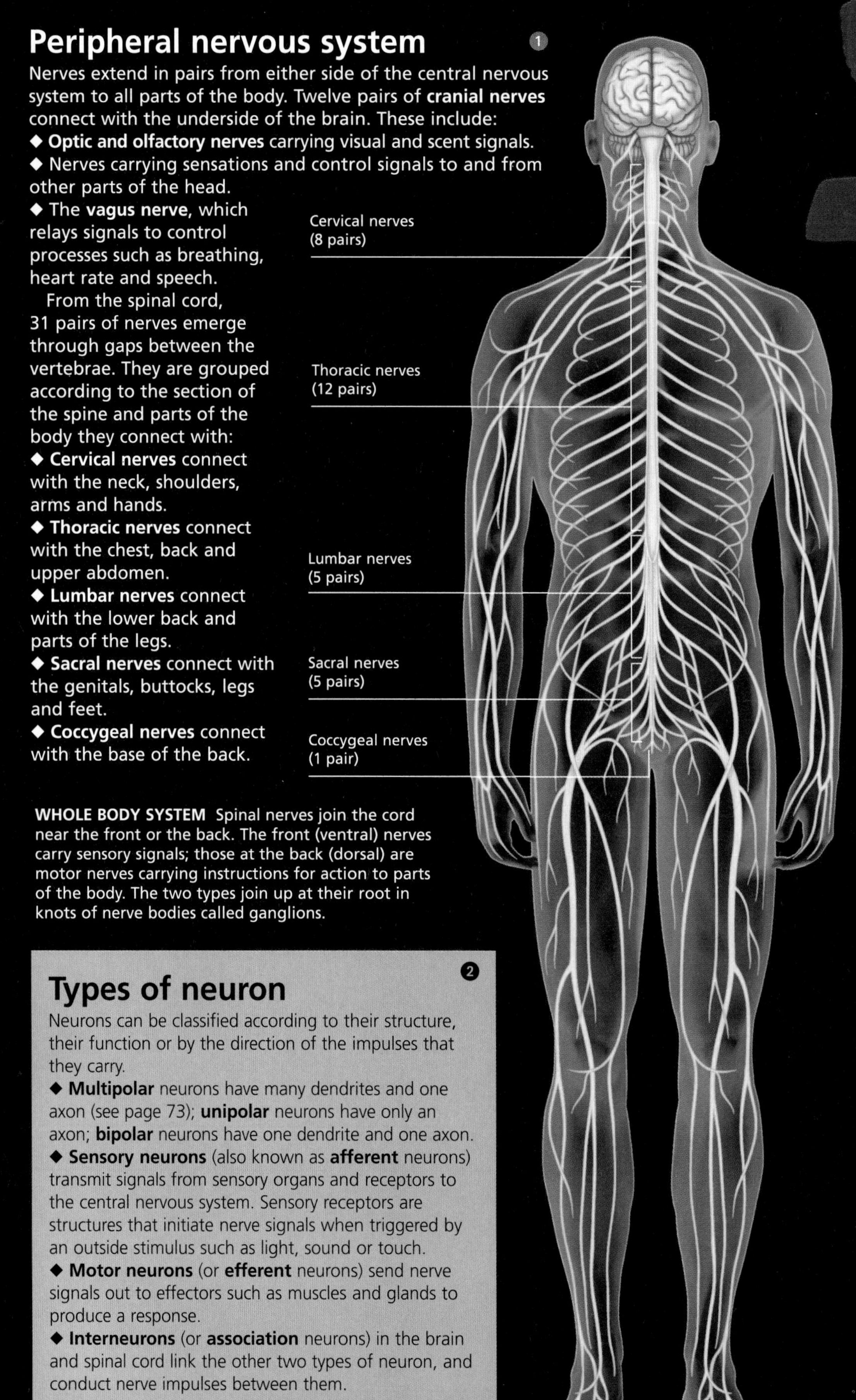

WHOLE BODY SYSTEM Spinal nerves join the cord near the front or the back. The front (ventral) nerves carry sensory signals; those at the back (dorsal) are motor nerves carrying instructions for action to parts of the body. The two types join up at their root in knots of nerve bodies called ganglions.

Types of neuron

2

Neurons can be classified according to their structure, their function or by the direction of the impulses that they carry.

◆ **Multipolar** neurons have many dendrites and one axon (see page 73); **unipolar** neurons have only an axon; **bipolar** neurons have one dendrite and one axon.
◆ **Sensory neurons** (also known as **afferent** neurons) transmit signals from sensory organs and receptors to the central nervous system. Sensory receptors are structures that initiate nerve signals when triggered by an outside stimulus such as light, sound or touch.
◆ **Motor neurons** (or **efferent** neurons) send nerve signals out to effectors such as muscles and glands to produce a response.
◆ **Interneurons** (or **association** neurons) in the brain and spinal cord link the other two types of neuron, and conduct nerve impulses between them.

WEIRD AND WONDERFUL

3

Neurons **damaged by injury** or surgery may, after some months, regrow along their original path, restoring feeling or movement. After a **stroke**, new nerve pathways may form in the brain to compensate for those lost.

Reflexes

5

If you touch a very hot object, the signal to your muscles to pull your hand away acts before the sensory signal conveying a burning sensation reaches your brain. This is an example of the simplest chain of cause and effect in the nervous system – a reflex.

Reflexes begin with the stimulation of a sensory receptor. A **sensory neuron** carries the impulse to the central nervous system, but an **interneuron**, usually in the spinal cord, triggers an impulse along a **motor neuron** to an effector, such as a muscle, provoking a response.

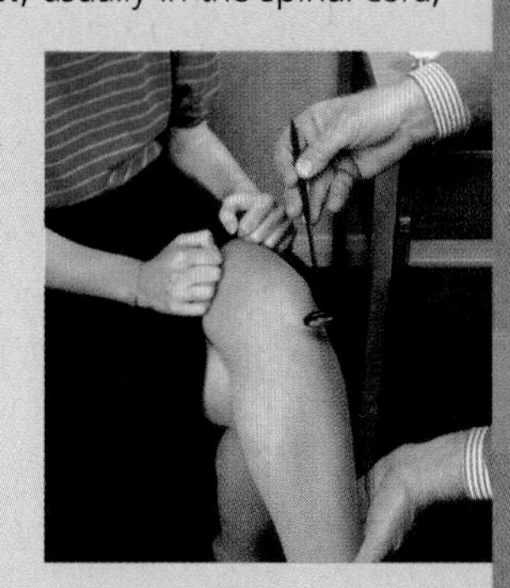

STRETCH REFLEX Doctors test the nervous system by tapping the knee, which should straighten the leg.

TIMESCALE

6

- ▶ **1543** Belgian Andreas Vesalius publishes the first fairly accurate drawings of the human nervous system.
- ▶ **1786** Italian Luigi Galvani attaches metal conductors to dead frogs during electrical storms. Their muscles twitch, showing that electrical signals play a role in muscle contraction.
- ▶ **1873** Italian Camillo Golgi stains nerve cells with silver nitrate to make them visible.
- ▶ **1880s** Spaniard Santiago Ramón y Cajal uses Golgi's technique to start working out the structure of the nervous system.
- ▶ **1924** German Hans Berger detects electrical signals in the brain.
- ▶ **1950s-60s** US neuroscientist Roger W. Sperry establishes that the brain's left and right hemispheres control different functions.
- ▶ **2000** Research into neurotransmitters by Arvid Carlsson, Paul Greengard and Eric Kandel leads to new treatments for neurological conditions.

Tie-breaker

4

Q At what age does the human spinal cord stop growing?

A Four years. The adult spinal cord is shorter than the vertebral column, finishing around the small of the back. In spinal anaesthesia, or lumbar puncture, the needle is introduced below the cord and so does not damage it.

Silent nervous system

Vital bodily activities such as breathing, heart rate, intestinal muscle contractions and glandular secretions are 'automatically' controlled by the **autonomic nervous system**. This involves two sets of nerves – **sympathetic** nerves, which usually speed up an activity or boost secretions, and **parasympathetic** nerves, which usually slow down an activity or reduce secretions.

SPINAL DIMENSIONS The adult spinal cord (the backtground image shows the cervical section) is approximately 42-45 cm (16-18 in) long. It is about 2.5 cm (1 in) in diameter, though it is thicker at the neck and also at the base of the back, where the nerves supplying the limbs meet the spine.

Major nervous system disorders

7

Central nervous system disorders cause symptoms ranging from headaches and vomiting to problems with neurological processing. Peripheral nerve damage may affect the senses, muscle movements or automatic control of parts of the body such as the bladder.

CNS disorders	Features
Alzheimer's disease	Degeneration of neural tissue in the brain.
Brain tumour	An abnormal mass; can distort nerves and disturb brain processes.
Epilepsy	Seizures caused by sudden electrical activity of nerve cells in the brain.
Meningitis	Inflammation of the membranes of the brain and/or spinal cord.
Creutzfeldt-Jakob disease (CJD)	Irreversible neuron damage by abnormal proteins called prions.
Stroke	Damage of brain tissue by blood vessel blockage or haemorrhage.

PNS disorders	Causes
Bell's palsy	Partial paralysis of the facial muscles caused by nerve damage.
Carpal tunnel syndrome	Pain in the fingers caused by ligaments pressing the wrist nerve.
Sciatica	Back pain from pressure on the sciatic nerve (main leg nerve).

QUESTION NUMBER

The numbers or star following the answers refer to information boxes on the right.

ANSWERS

21 True ❻

30 True ❶ ❷

94 Pupil ❶ ❷

195 False ❻

294 Red and green ❻

298 Yellow spot ❷

322 The cornea ❷ ❹

415 **'[They] only have eyes for you'** – written in 1934

419 **Suzie** – released in 1972

431 **1930s** – in 1938

475 Glaucoma ❹

500 **True** (why older people need better lighting) ❷

506 (a) Vitreous, (b) Aqueous ❷

★ **524** C: A crossing or intersection

698 **Claude Rains** – based on H.G. Wells novel

724 Stereoscopic ★

762 Retina ❶ ❷

789 The eyes ❷

902 Reading glasses ❸

947 ***The Eyes of Laura Mars*** – a murder thriller

954 D: Optic disc ❷

990 **Van Morrison** – album released 1998

Eyes and vision

Core facts ❶

◆ The eyes are our most precise sensory organs, containing more than 70 per cent of the body's **sensory receptors**.

◆ The eye works like a **camera**, focusing rays of light to create an image. The curved **cornea** at the front bends light rays to a focus on the inner surface – the **retina**. The **lens** adjusts the focus, allowing near and far objects to be seen clearly. The **iris** adjusts the opening (the **pupil**) to allow in more or less light.

◆ The **shape** of the eyeball determines the eye's ability to **focus** sharply.

◆ The **distance** between the two eyes enables the brain to judge distance and dimension.

The eye ❷

The eyeball is roughly spherical, about 2.5 cm (1 in) in diameter, and filled with transparent, jelly-like fluid. Each eyeball is anchored into its bony socket, and rotated, by six small muscles.

The light-sensitive inner surface of the eyeball, the **retina**, contains two types of light receptor:

◆ Narrow **rod cells** which detect black and white and are active in dim light.

◆ Shorter, thicker **cone cells** which detect colours, but only operate in good light.

Cones are clustered in the centre of the retina at the back of the eye, particularly around the **fovea** – the point of sharpest focus.

★ 524 Nerve pathways ★

Signals from the optic nerves of each eye partly cross over at a junction in the brain called the **optic chiasma**. Signals from the right-hand field of view reach the left side of the visual cortex, and vice versa. The brain combines the two viewpoints to perceive distance and three dimensions – known as **stereoscopic vision**.

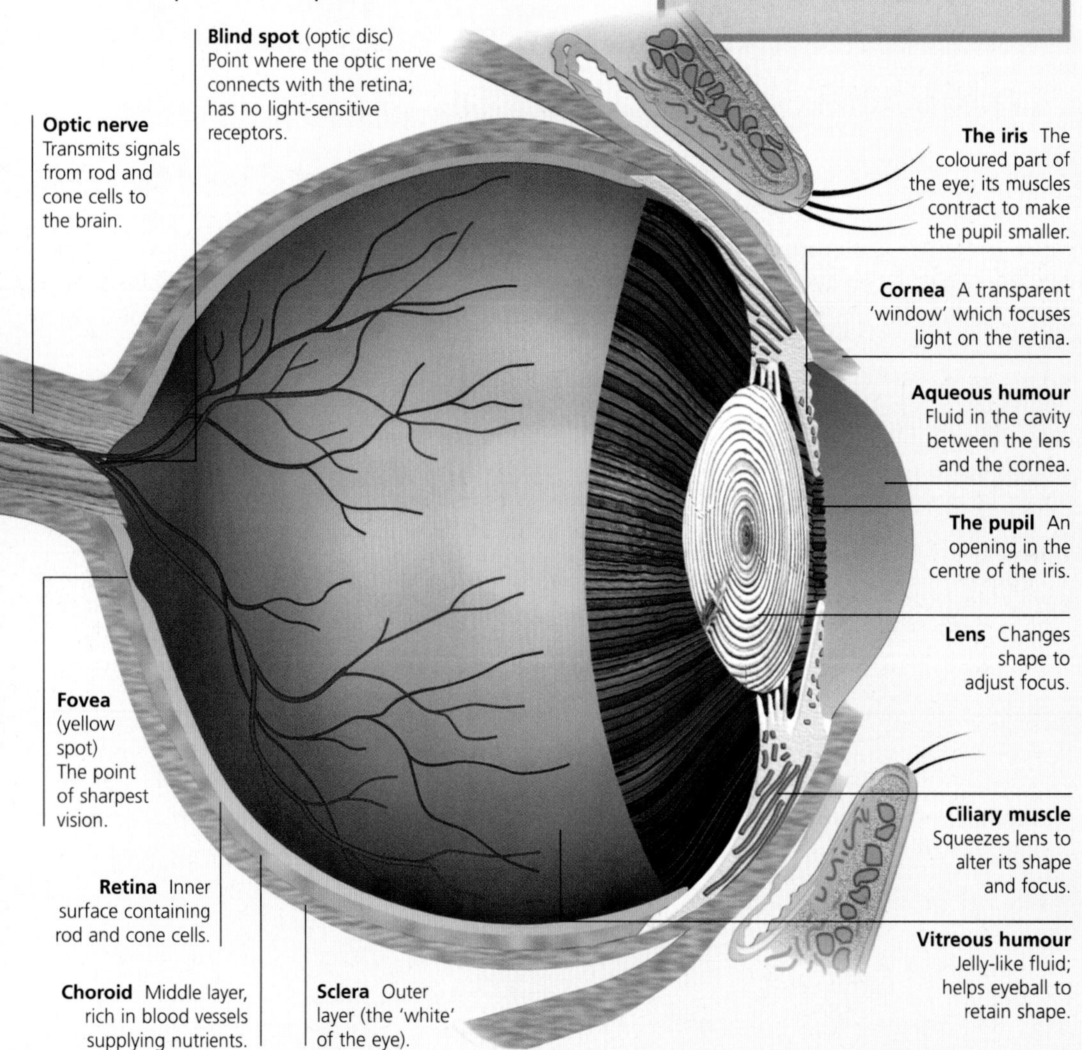

Focusing

3

In a process called **accommodation**, the convex cornea bends light rays from the visual field as they enter the eye and the lens adjusts the rays to focus a sharp image on the retina.

Disorders include **presbyopia**, in which the lens becomes stiff with age and cannot focus on nearby objects, and **astigmatism**, when the eye cannot focus sharply because the eyeball is not spherical.

HYPERMETROPIA In long-sightedness, images focus behind the retina, blurring nearby objects. This can be corrected with a convergent (convex) lens.

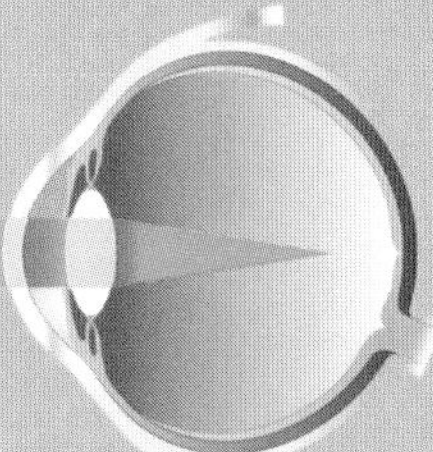

MYOPIA In short-sightedness, images focus in front of the retina, causing distant objects to blur. This is corrected with a divergent (concave) lens.

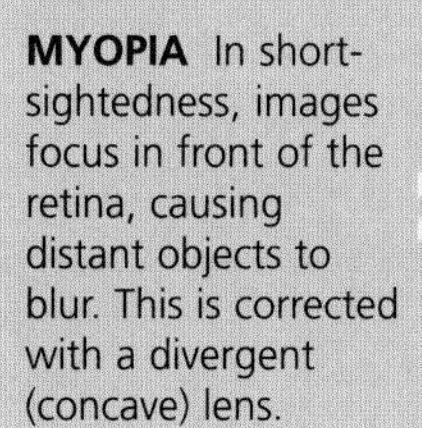

WEIRD AND WONDERFUL

5

Although the eye works in a similar way to a camera, there is a big difference between the two. Camera film cannot adjust its sensitivity, but the eyes adapt to dim light to let you see objects you could not, at first, make out.

Defective colour vision

6

Colour blindness can be caused by disease damaging the cone cells responsible for detecting either red, blue or green light. More often, however, it is caused by a **genetically inherited defect** in the cone cells. It causes differences in colours to be indistinct – strong colours are often more easily distinguishable than subtle shades.

The most common form is an inability to distinguish between red and green – a recessive condition inherited through the **X chromosome**. Men have one X chromosome – if the chromosome contains the defect, they will be colour blind. Women have two X chromosomes, so are more likely to be carriers. The hereditary condition affects 7 per cent of men and 0.5 per cent of women.

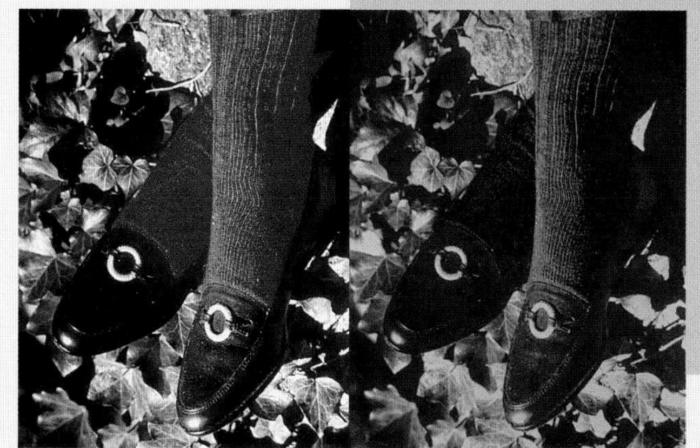

COLOUR CONFUSION Someone with red-green colour blindness, known as protanopia, may see red socks (far left) as almost black (left). It can also cause confusion between brown and orange, and different reds may be indistinct.

Eye diseases and disorders

4

Eye conditions resulting in partial or total blindness can be caused by poor nutrition, disease, genetic disorders, injury, infection or the degeneration of the tissues with age.

Disorder	Features	Treatment
Glaucoma	Build-up of aqueous humour causes high fluid pressure in eye.	May be treatable with drugs, or by laser or conventional surgery to open a drainage passage through or beside the iris.
Macular degeneration	Deterioration of the central area of the retina.	There is currently no treatment for this condition, which affects mainly people aged over 70.
Retinal detachment	Detachment of retina from choroid, caused by degeneration or injury.	May be correctable by laser surgery, if caught early, or by conventional surgery.
Trachoma	Bacterial infection causing scarring of the cornea.	Antibiotic tablets and eye drops. Common infection in developing countries.
Retinitis pigmentosa	Night blindness and tunnel vision; patches of dark pigment in eye.	This disease is inherited, and currently not treatable.

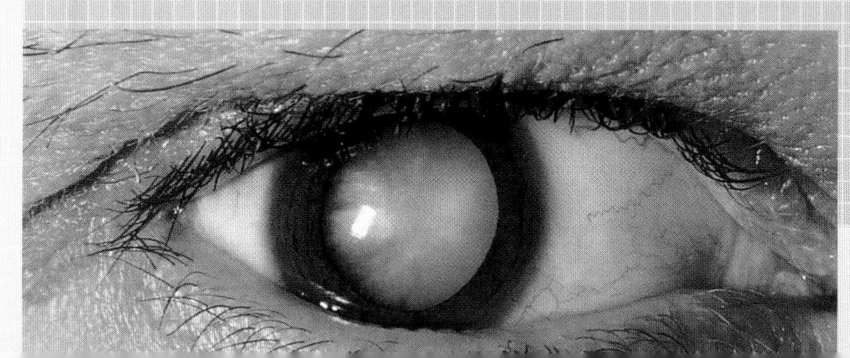

WHITE SHROUD A lack of nutrients reaching the lens can cause a build-up of proteins. This may lead to a cataract – a cloudy, opaque lens, resulting in poor vision.

QUICK CURE Cataracts can be treated with laser surgery (above). A laser beam breaks up the lens, which is then removed and replaced with a clear plastic lens.

QUESTION NUMBER

The numbers or star following the answers refer to information boxes on the right.

ANSWERS

38	Cochlea ❷
100	Sound, or hearing ❶ ❷
★ 113	Hypercusis
161	In the ear (parts of the balance apparatus) ❷ ❹
162	Movement (specifically, turning) ❹
163	Frequency of sound waves – determines pitch
164	Amplify it ❷
165	1000 times (the decibel scale is logarithmic) ❻
236	1950s ❺
273	Ear trumpet ❺
394	Hammer, anvil and stirrup ❷
518	Eustachian tube ❷
658	Air pressure (on each side of eardrum) ❷
702	Tympanum ❷
703	The labyrinth ❷
796	Beethoven (he became deaf) ❺
913	Tinnitus ❺
914	Profound deafness ❺
986	Stevie Wonder – on 'Hotter Than July', released 1980

Ears, hearing and balance

Core facts ❶

◆ Sounds consist of waves or vibrations of air molecules. The ear's external flap, the **pinna**, funnels these vibrations into the **ear canal**. Together these make up the **outer ear**.

◆ The **middle ear**, beyond the eardrum, transmits vibrations to the **inner ear**, where sounds are sensed and auditory (sound) nerve signals are sent to the brain. These parts are all embedded in a cavity in the skull.

◆ Having two ears enables the listener to sense the direction of sounds – **stereophony**.

◆ The inner ear has a key role in **balance**. It signals to the brain which way up the head is, and if it moves or rotates.

Inside the ear ❷

Sound vibrations or waves make the **eardrum** vibrate, then pass into the middle ear, which has three tiny bones. The **hammer** (maleus) vibrates with the eardrum; the **anvil** (incus) strengthens these vibrations, and the **stirrup** (stapes) passes the strengthened vibrations into the inner ear.

The inner ear is a complex series of fluid-filled channels sometimes called the **labyrinth**. They consist of the **vestibule** and **semicircular canals** (concerned with balance) and the spiral **cochlea**. The sensory receptors for sound are in the **organ of Corti** inside the cochlea. This responds to sound vibrations and sends nerve signals to the auditory part of the brain.

WEIRD AND WONDERFUL ❸

To hear, your brain must interpret the sounds your ear receives. Only three sounds are interpreted in the same way by everyone: a baby crying, an alarm clock ringing, and hearing your name spoken all trigger the fight or flight response.

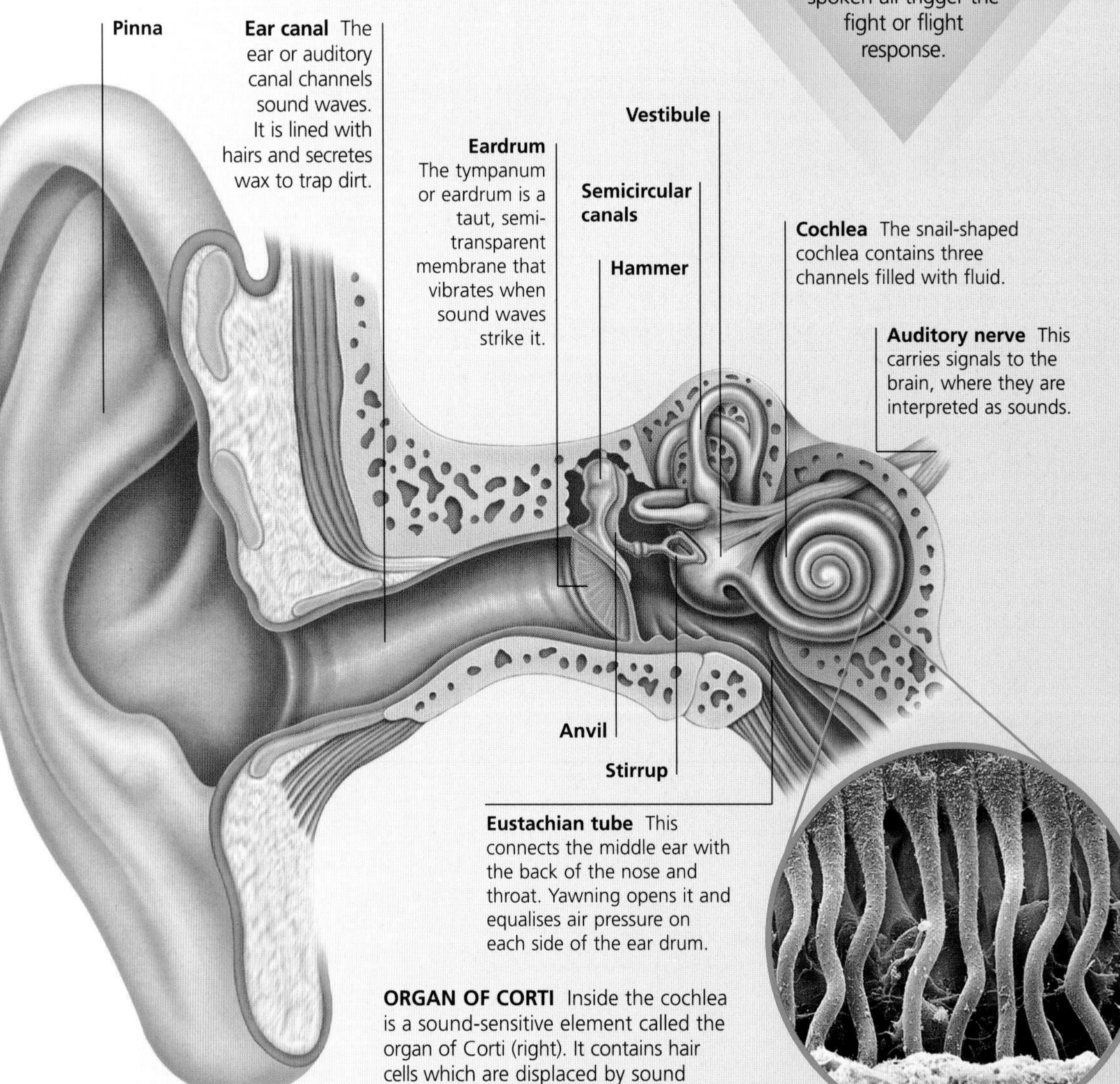

ORGAN OF CORTI Inside the cochlea is a sound-sensitive element called the organ of Corti (right). It contains hair cells which are displaced by sound waves. This stimulates nerve signals, which are then transmitted to the brain.

Balance

The vestibule and semicircular canals are the parts of the inner ear that detect the position and movement of the head.

The **vestibule** contains two chambers – the **utricle** and **saccule** – which respond to gravity and signal to the brain which way up the head is. They have a soft membrane and tiny sensory hairs, and work by detecting the displacement of the hairs when the membrane is pulled on and distorted by gravity.

The **semicircular canals** are arranged at right-angles in three dimensions. They also have sensory hairs. The hairs do not move if the head remains still or moves steadily. But if it turns, the fluid in one or more of the canals swirls and deflects the hairs. This triggers signals to the brain. Spinning round and round quickly, then stopping, makes the world seem to continue turning because the fluid continues to swirl for a few moments.

GRAVITY SENSOR Microscopic hair cells in the vestibule of the inner ear detect gravity. They report to the brain on the direction and strength of body movements – so that it knows whether you are upside-down, or accelerating.

113

Ear and brain

The brain heightens or suppresses noise according to how you perceive it. For example, if a neighbour's loud music annoys you, you become hyper-aware of it and may suffer stress symptoms. This is called **hypercusis**. But if you do not find it intrusive, your brain interprets it as background noise and you hear it less acutely. This is called **hypacusia**.

Disorders of the ear

- Partial or total **deafness** may be caused by ageing, injury, disease, an inherited defect or the overgrowth of bone in the middle ear. **Hearing aids** help deafness in various ways. The most common ones boost the sound entering the ear. For people whose middle ear cannot transmit airborne sounds, there are aids that transmit sounds to the inner ear through the skull, and cochlea implants, which stimulate the auditory nerve.
- Anything that affects the inner ear may cause **dizziness**, often with nausea. Problems with the inner ear may also cause **tinnitus** (ringing, hissing or other noises in the ears). A serious but rare disorder is **Ménière's disease**, in which the fluid pressure in the inner ear rises. **Labyrinthitis** – an inner ear inflammation, often caused by a viral infection – is more common, and often clears up by itself.
- **Motion sickness** causes symptoms such as dizziness and nausea in many people when travelling. This may be because the brain is receiving contradictory messages from the eyes and the balance organs of the ears.

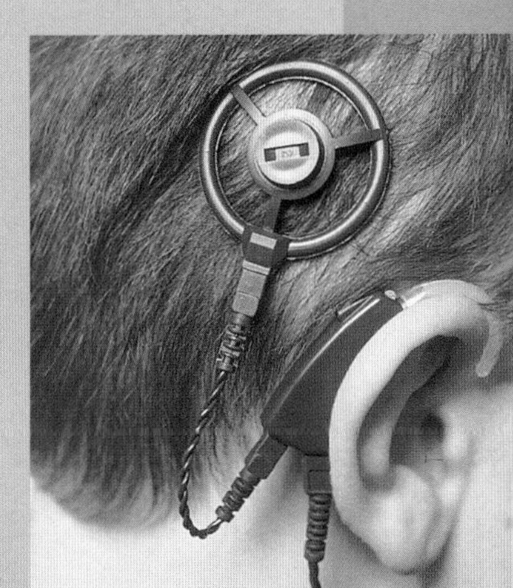

COCHLEAR IMPLANT A processor transmits sounds to the implant. This converts them into signals which an electrode transmits direct to the auditory nerve. They are recognised by the brain as sound.

Decibels

Sound scale	Equivalent to:
0	Faintest audible sound
10	Rustling leaves; quiet whisper
20	Average whisper
30-40	Quiet conversation
50	Normal speech
60-65	Loud conversation in busy office or pub
65-70	Traffic on average street
65-90	Train; bus or lorry interior
75-80	Light industrial factory
90	Motorway; other heavy traffic
90-100	Electric saw; nearby thunder
110	Loud orchestral music (in audience)
120	Loud rock music (in audience)
130	Threshhold of ear pain
130-140	Nearby artillery fire; jet engine
up to 190	Space rocket taking off

QUESTION NUMBER

The numbers or star following the answers refer to information boxes on the right.

ANSWERS

114 ***Remembrance of Things Past*** – published 1913-27

146 ***The Scent of a Woman*** – Pacino won an Oscar

167 **Cold, heat, pain, pressure, vibration** ❶ ❷

168 **Smell and taste** ❶ ❸

169 **Monosodium glutamate (MSG)** – rich and meaty

170 **Taste** ❺

200 **False** ❹

290 **True** ❷

369 **Its smell** – it is the smelliest known chemical

519 **Taste buds** ❹

689 **True** ❸

693 **Lear** – by William Shakespeare

697 **Rita Tushingham** – directed by Tony Richardson

★ **755** **Counter-irritation**

783 **Mouth and nasal cavity or nose** ❸

814 **Touch** ❷

921 **Rhinitis** – can be caused by a cold or an allergy

Touch, taste and smell

Core facts ❶

◆ The **sense of touch** consists of four related senses: cold, heat, pain and touch itself (which also detects pressure and vibration). Touch receptors are found all over the body.
◆ The **senses of taste** (gustation) and **smell** (olfaction) are closely related. They are 'chemical senses' because they detect chemical stimuli.
◆ The **flavour of food and drink** is really a combination of taste, smell and mouth-touch.
◆ These senses exist throughout the animal kingdom. Many animals have a much more highly developed sense of smell than humans, and their brain's olfactory 'bulbs', which process smell signals, are proportionately far bigger.

Touch receptors ❷

Different types of receptors – in the skin and other parts of the body – respond to different kinds of touch. The simplest are **free nerve endings**, which send signals that the brain interprets as simple touch, temperature, pressure or pain. More specialised receptors include **Merkel's discs** and **Meissner's corpuscles** (so named after their discoverers), which are sensitive to light touch; **Krause's end-bulbs** and **Pacinian corpuscles** which are sensitive to vibrations and deep pressure; and **Ruffini corpuscles**, sensitive to stretching.

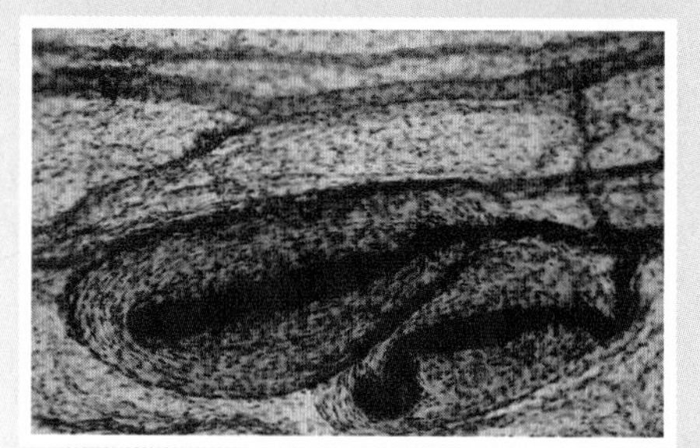

PACINIAN CORPUSCLES These respond to fine stimulation and low-frequency vibration within a second of contact.

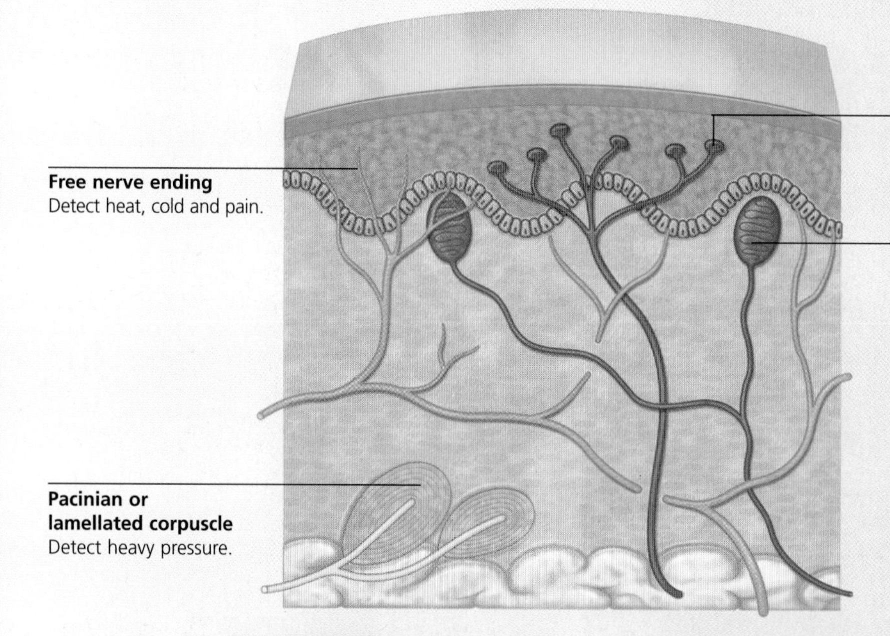

★ 755

Itching and scratching

Itching is a symptom of many skin conditions, some trivial, some not. Soothing creams and ointments may help with itching, but there is often an urge to **scratch**. This is known as **counter-irritation**. When we scratch we stimulate nerve fibres near those causing the itch so strongly that – temporarily, at least – the itching sensation is drowned out.

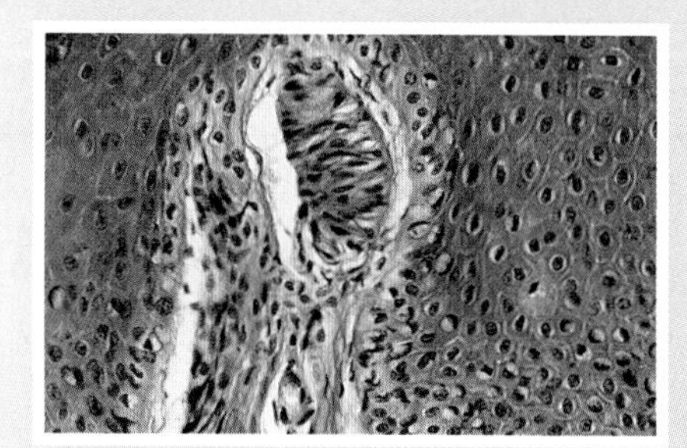

MEISSNER'S CORPUSCLES These light-touch receptors are in the fingertips, lips and other sensitive areas of the body.

Anatomy of taste and smell

❸

Sensory receptors in the nose and tongue respond to the **chemical nature** of food, drink and airborne materials (which must first dissolve in nasal mucus). The throat (pharynx) links the nose and mouth, so we usually taste and smell things at the same time. This is why a blocked nose makes things such as coffee and chocolate lose much of their flavour. On the other hand, some chemical 'smells', such as that of chloroform, are in fact tasted, not smelled. Smell is one of the most **primitive and basic of the senses**, and is closely linked to the parts of the brain involved in memory – probably the reason why smells can evoke vivid memories.

WEIRD AND WONDERFUL

❺

One of the nerves taking taste signals from the tongue to the brain is the **chorda tympani**, which crosses part of the ear drum on its way to the brain. As a result, injury to the ear drum can affect the sense of taste.

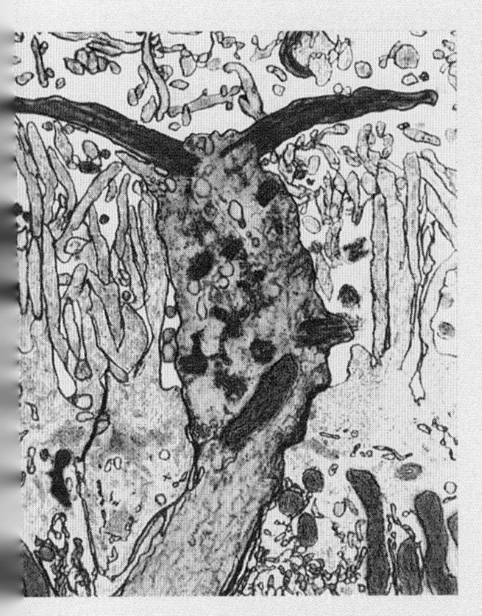

SMELL DETECTOR A smell receptor in the nose's lining. The two hair-like 'cilia' at the top are thought to transmit smells from the air to the cell below.

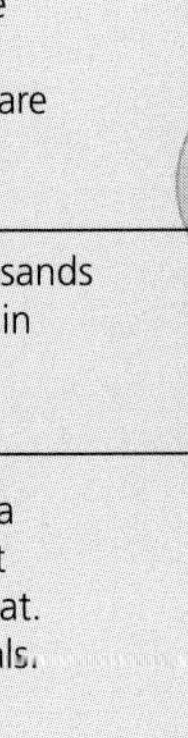

Nasal cavity Lined with mucus and millions of olfactory hairs – receptors sensitive to chemical particles in the air breathed in and dissolved in the mucus.

Pharynx A passage connecting the back of the nose to the trachea (windpipe) and the oesophagus (food tube). The upper part, the nasopharynx, runs from the nasal cavities to the area behind the soft palate. Taste receptors are distributed all over it.

Tongue This has thousands of taste buds recessed in pores on its surface.

Salivary glands Saliva keeps the mouth moist and lubricates the throat. Dissolves food chemicals.

Taste buds

❹

Each taste bud is an onion-shaped cluster of 50-100 receptor cells. They are scattered unevenly over the tongue, with fewer in the centre. An average adult has about 3000 – some taste-sensitive people have as many as 10 000. All buds respond to all tastes, but the brain interprets their signals as different tastes, according to the pattern of stimulation.

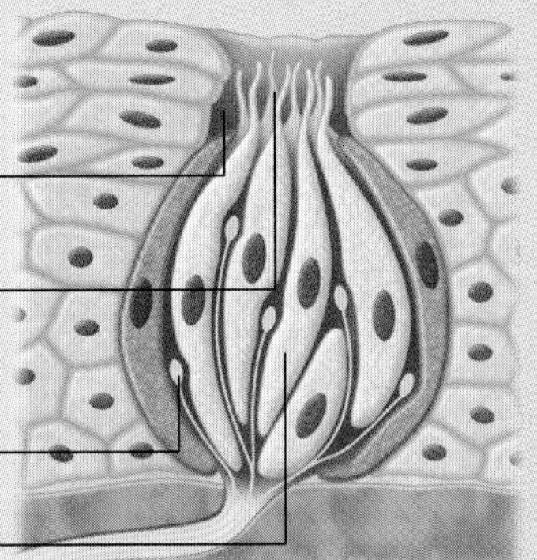

Taste pore A narrow opening onto the skin surface.

Taste hairs Extend into the saliva in the mouth.

Gustatory (taste) cells Stimulated by chemicals dissolved in the saliva contacting the taste hairs.

Supporting cells Surround gustatory cells.

TASTE BUDS These are receptors that send taste sensations to the brain.

The strength of smells

❻

How strong smells are is not the same as how pleasant or unpleasant they may be. Vanilla smells sweet but far **smaller amounts of it can be detected** than of rotten eggs. Here is how a few smells compare for potency – for example, the smell of rose oil is 50 times as strong as that of rotten eggs:

Smell	Potency	Smell	Potency
Rotten eggs	1	Rose oil	50
Citrus peel	2	Garlic oil	300
Almond essence	25	Vanilla	30 000

Glands and hormones

QUESTION NUMBER

The numbers or star following the answers refer to information boxes on the right.

ANSWERS

7 **Adrenaline** ❸ ❹

31 **Insulin** ★

209 **Insulin** ❼ ★

323 **Growth hormone** ❸ ❺

370 **The thyroid** ❷ ❸

507 **Into the bloodstream** ❶ ❷

508 **The islets of Langerhans** ❷ ❸

520 **Endocrine glands** ❶ ❷

624 **Adrenal glands** (Latin *ad* = near, *renes* = kidneys) ❷

679 **The pituitary** ❷

704 **Hormone system** ❶ ❷

★ 815 **Diabetes**

825 **Adrenal** (specifically, the adrenal cortex) ❼

955 **A: Calcium** ❷ ❸

Core facts ❶

◆ Glands secrete chemical substances called **hormones**. Hormones act as chemical messengers, carrying signals to target organs in other parts of the body to modify their activity.

◆ The body contains **endocrine** (ductless) glands, which secrete hormones into the blood, and **exocrine** (ducted) glands, such as the liver and the sweat and salivary glands, which secrete substances that are expelled from the body.

◆ The release of hormones is connected to the nervous system through the **hypothalamus**.

◆ Hormones maintain homeostatis (balance) in the body, regulating processes such as **food metabolism**, **growth** and **reproduction**.

The endocrine system ❷

The endocrine system consists of **glands** which secrete **hormones** (chemicals made from proteins or lipids) into the blood. Hormones travel to target tissues and cause changes in their activity to maintain the **physiological balance** of the body (for example, regulating blood glucose levels). The system is controlled in the brain by the **hypothalamus**, which monitors hormone levels in the blood and stimulates or inhibits the **pituitary gland**. The pituitary regulates the secretions of the other glands in the endocrine system.

AUTOMATIC CONTROL
The endocrine system operates a 'feedback' mechanism to ensure that the blood contains appropriate hormone levels. Excess blood levels of a particular hormone send a signal to the secreting gland which inhibits further production.

Pineal gland Regulates sleeping and reproductive functions.

Hypothalamus The 'control centre' of the endocrine system.

Pituitary gland The 'master gland', controlled by the hypothalamus.

Thyroid gland Controls the body's metabolic rate. The **parathyroids** – four small glands behind the thyroid – help to regulate calcium levels.

Adrenals Two glands just above the kidneys which secrete hormones involved in the stress response.

Pancreas Groups of pancreatic cells – the 'islets of Langerhans' – produce hormones that regulate blood sugar.

Ovaries Produce female sex hormones.

Testes Produce male sex hormones.

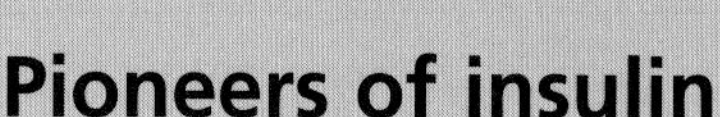

★

Pioneers of insulin

In 1921-2, Canadian doctor Frederick Banting and his team, working in the laboratory of John MacLeod, became the first scientists to extract and isolate **insulin** – the hormone responsible for blood-sugar regulation. In 1922 Banting used it to treat diabetes. For decades, insulin was extracted from the pancreas of animals. Today it is created by **genetically engineered** bacteria.

BACKGROUND IMAGE Magnified crystals of the female hormone progesterone.

Important hormones

Hormone	Target	Function
Pituitary gland		
Adrenocorticotrophic hormone	Adrenal cortex	Stimulates release of cortisone.
Follicle-stimulating hormone	Ovaries/testes	Stimulates egg/sperm production.
Growth hormone	All tissues	Protein and fat metabolism; growth.
Thyroid-stimulating hormone	Thyroid	Stimulates release of hormones.
Pineal gland		
Melatonin	Hypothalamus	Regulates sleep/reproduction.
Thyroid gland		
Calcitonin	Bones	Lowers excess calcium in body.
Thyroxine and tri-iodothyronine	Cells	Stimulate metabolism.
Adrenal glands		
Aldosterone	Kidneys	Balances body fluid levels.
Cortisone and hydrocortisone	Most tissues	Protein and fat metabolism; glucose production; inhibit inflammation.
Pancreas (islets of Langerhans)		
Glucagon and insulin	Liver; muscles	Regulate blood sugar levels.
Ovaries and testes		
Sex hormones oestrogen and progesterone (ovaries) and testosterone (testes).	Sex organs; pituitary gland	Promote development of sexual characteristics (oestrogen/testosterone); regulate ovulation/pregnancy in women.

BULL RUN The fight or flight response was vital to our ancestors, and our survival sometimes still depends on it.

Fight or flight

In a threatening situation, the nervous system triggers the release of **adrenaline** into the bloodstream. This increases heart and breathing rates, stimulates sugar release from the liver for energy and diverts blood to the muscles, preparing the body to fight or run away. This response has not changed since prehistoric times, but modern stresses do not always allow fight or flight. In this situation, excessive adrenaline can lead to **high blood pressure** and **heart problems**.

Gigantism and dwarfism

The hypersecretion (overproduction) of growth hormone (GH) in childhood, often associated with a pituitary tumour, causes **pituitary gigantism**. The skeleton lengthens with excessive cartilage and bone, and sufferers may grow to 2.4 m (8 ft) tall. Overproduction during adulthood (**acromegaly**) causes new bone and cartilage growth, mostly affecting the face and hands.

Underproduction (hyposecretion) of GH in childhood leads to **pituitary dwarfism**, causing a person to be small but of normal proportions – this is not the same as the genetic condition **achondroplasia**, which leads to a lack of cartilage cells and produces short limbs and an enlarged forehead.

SMALL PROPORTIONS The endocrine system can disrupt the usual pattern of growth.

WEIRD AND WONDERFUL

When a baby is ready to be born, its endocrine system is thought to secrete **adrenal hormones** into the placental blood supply, changing the mother's hormone levels and making the uterus (womb) irritable.

Major disorders

- **Thyroid problems** An excess of thyroxine – hyperthyroidism – raises the metabolic rate. Symptoms include anxiety, heat intolerance and palpitations. Hypothyroidism – low levels of thyroxine – causes low energy and intolerance of the cold. Both can arise from diseases such as auto-immune disorders.
- **Adrenal problems** Damage to the adrenal cortex leads to Addison's disease, a deficiency of adrenocortical hormones, with symptoms such as weakness, nausea and weight loss. Adrenal or pituitary tumours can cause Cushing's syndrome – excess hydrocortisone, leading to weight gain on the face, shoulders and pelvis, and osteoporosis.
- **Pancreatic problems** Damage to the pancreas can cause diabetes mellitus – insulin deficiency, leading to a rise in blood sugar. Symptoms include excessive thirst and urination, and weight loss.

Sex and reproduction 1

QUESTION NUMBER

The numbers or star following the answers refer to information boxes on the right.

ANSWERS

46 **R18** – R for restricted

90 **Sexually transmitted infection** ❼

99 **Cervix** ❷ ❼

193 **True** ❸ ❹

212 **Testosterone and oestrogen** ❶ ❷ ❺

214 **Adds fluid to semen** (to activate sperm) ❷ ❼

215 **The penis** ❶ ❷ ❻ ❼

265 **The larynx, or voice box** ❺

295 **Yellow** – *luteum* in Latin

422 **Pheromones** ❻

538 **Fallopian tubes** ❷

746 **Fall in love** ❻

786 **The prostate** ❷ ❼

★ 856 **True**

916 **Warts** ❼

943 ***Sex*** – directed by Steven Soderbergh

Core facts ❶

◆ A man's testes make **sperm**, and woman's ovaries **eggs**. The reproductive organs also make the **sex hormones** that cause the changes of puberty.
◆ Men make huge numbers of sperm all the time. Women ovulate (produce an egg) once in a **menstrual cycle** – roughly every 28 days.
◆ Sperm ejaculated from a man's penis into a woman's vagina during sex swim up her reproductive tract in order to meet an egg.
◆ Out of millions of sperm ejaculated, only one can **fertilise** each egg. **Multiple births** result from more than one egg being fertilised or from a fertilised egg later splitting in two.

Reproductive systems ❷

From puberty to the menopause, if a woman is not taking contraceptive hormones, her body **ovulates** (releases an egg) as part of the menstrual cycle, roughly every 28 days.

Unlike other mammals, however, human females do not signal the time of ovulation to males by obvious bodily changes, such as in the sexual organs, so it is not obvious when this time of greatest fertility occurs. It is controlled by the pituitary gland in the brain.

Men do not have hormonal cycles. Sperm production is regulated by **gonadotrophins** – hormones that are produced by the testes when stimulated by a hormone from the hypothalamus gland. Gonadotrophins stimulate the testes to produce **testosterone**, the male hormone.

MALE GENITALS Their task is simply to make sperm (in the testes) and the seminal fluid in which sperm swims (secreted by glands), then deliver it during ejaculation. Some parts are shared with the urinary system.

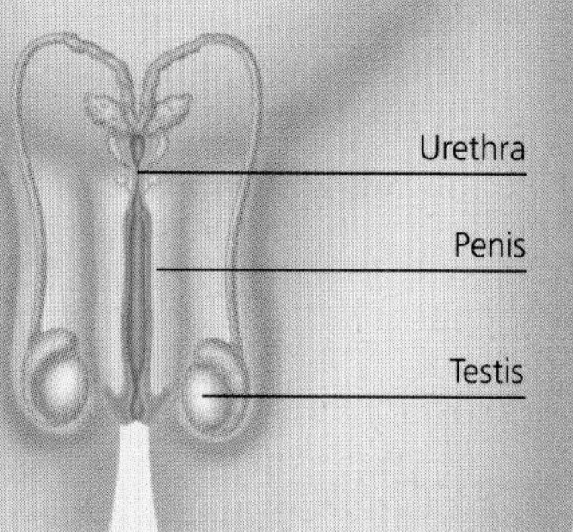

FEMALE GENITALS These store and mature eggs, provide a place for fertilisation, then shelter and nourish the developing foetus until it is ready to be born.

Fallopian tube
Ovary
Womb (uterus)
Cervix
Vagina

OVUM ON THE MOVE From puberty, one egg matures almost every month and is released from its ovary into the adjoining fallopian tube to start its journey to the womb.

★ 856

Cooler for fertility

Sperm develop best at about 2°C (3.6°F) lower than the normal body temperature of 37°C (98.6°F). Having the **testes** hanging just outside the body, in the **scrotum**, keeps them just that much cooler. Tight underwear and excess fat in the lower body may reduce the number of sperm produced.

WEIRD AND WONDERFUL ❸

Until 1930, women were thought to ovulate at the time of their periods. Then researchers in Japan and Austria found that **ovulation** takes place about mid-way between periods – when they were thought to be least fertile.

Menstrual cycle ❹

Day 5 Oestrogen causes the womb lining to regrow. Follicle-stimulating hormone (FSH) causes an egg to ripen.
Day 11 Luteinising hormone (LH) is released, triggering ovulation.
Day 21 Progesterone prepares the womb lining to receive a fertilised egg. If this doesn't happen it diminishes, triggering a period.

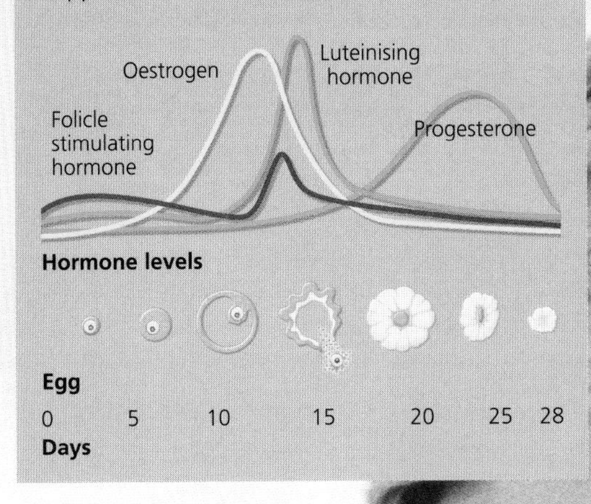

POSITIVE ATTRACTION Most human senses can be involved in sexual attraction, leading to the instinct to mate and reproduce.

Sexual attraction and the senses ❻

Initial attraction is often **visual**, but psychologists argue about the significance of specific visual cues, such as the face, legs, bottom and a woman's breasts, and whether they signal desirable attributes in a mate or are conditioned by fashion. In many animals, smelling substances called **pheromones** are important, and the same may be true of humans – yet perfumes and deodorants mask such natural scents. At closer proximity, the **sense of touch** becomes the most important – in the subtle stimulation of the body's so-called erogenous zones, which heightens **sexual arousal**, and later in the direct stimulation of nerve endings in the penis, clitoris and vagina that brings orgasm and ejaculation.

Changes at puberty ❺

During puberty, the body matures sexually. The process normally begins between the ages of 9 and 14. The **pituitary gland** at the base of the brain produces **growth hormones** that stimulate a growth spurt; and **follicle-stimulating hormone** (FSH) and **luteinising hormone** (LH), which control the development of sex organs and the production of the sex hormones, oestrogen (female) and testosterone (male).

BOY TO MAN For boys, one of the most obvious changes in puberty is the deepening of the voice known as the voice 'breaking'. Boy trebles, like Aled Jones, must wait to see how their singing is affected.

Major diseases ❼

◆ **Cancer** is a significant problem in the reproductive systems of both sexes. It includes, in men, cancer of the testes (one of the most common cancers in men under 40), the prostate (the most common male cancer) and, more rarely, the penis. Women may have cancer of the ovaries, uterus, cervix, breasts or, more rarely, vagina and/or vulva. Breast and cervical cancer are among the most common cancers in women.

◆ **Sexually transmitted infections** (STIs) affect and are passed between both sexes. The most dangerous is HIV, the cause of AIDS. Others are chlamydia (which can lead to infertility) genital herpes, gonorrhoea, syphilis and trichomoniasis. Genital warts are common and painless, but dangerous because the virus that causes them is also the main cause of cervical cancer in women. In women, a sexually transmitted disease may lead to more widespread infection known as pelvic infammatory disease (PID).

◆ **Other common problems** in women include endometriosis, in which pieces of uterine lining become attached to other internal organs and cause bleeding in time with the menstrual cycle; ovarian cysts, sometimes cancerous; prolapse of the uterus, in which it protrudes into the vagina; and vaginal thrush, a fungal infection. In men problems include an enlarged (but not cancerous) prostate – this mostly happens to older men.

QUESTION NUMBER

The numbers or star following the answers refer to information boxes on the right.

ANSWERS

QUESTION NUMBER	ANSWERS
20	**Foetus** 4
85	**Intrauterine device** 2
86	**In-vitro fertilisation** 6
132	**B: 25 mm (1 in)** 4
140	**D: 25 000 times** 1
154	**IUD** – others are forms of assisted conception 2
213	**Egg, or ovum** 3
★ 216	**(a) Female, (b) Male**
384	**True** 5
385	**False** 3
390	**False** (it develops from the embryo) 1
408	**D: 250 million** 3
409	**C: 200 000** 3
462	**Rubella, or German measles** – a viral disease
467	**High blood pressure** 7
482	**Siam** (they were the original 'Siamese' twins) 5
915	**Ectopic pregnancy** 7
936	**False** (only one sperm enters) 1
950	***Maybe Baby*** – written and directed by Ben Elton
971	**Uterus** 1

Sex and reproduction 2

Fertilisation 1

Fertilisation takes place high in one of the **Fallopian tubes**, the ducts that connect the ovaries with the **womb** (or uterus) in the female reproductive tract. The **egg** (or ovum) comes down the Fallopian tube from the ovary. If sexual intercourse has taken place, it meets millions of **tadpole-shaped sperm cells** (little more than a nucleus with a tail), swimming up the reproductive tract. The sperm cells cluster around an egg, some 25 000 times their bulk, and eventually one of them penetrates the egg's outer membrane with its 'head'. The sperm cell's nucleus discards its tail and merges with the nucleus of the egg. The **fertilised egg**, known as a **zygote**, then starts to travel down the Fallopian tube to the womb. It splits first into two cells, then into a cluster of cells, which will embed itself in the lining of the womb.

Womb (uterus)

Ovary

Fallopian tube

ZYGOTE The fertilised egg (above) divides into two cells within about 30 hours of fertilisation.

MORULA During the first week, the zygote (left) divides many times to form a cluster of cells, known as a morula.

BLASTOCYST By six to eight days after fertilisation, the egg (left) has divided into 100 cells. The cluster (now called a blastocyst) embeds itself in the womb's lining, forming what will become the placenta.

WEIRD AND WONDERFUL 3

A girl is born with all her eggs – some 200 000 – in immature form as oocytes. About 500 will eventually mature and be released. The testes make about 125 million sperm a day. In a typical ejaculation about 250 million are released.

Contraception 2

◆ **Barrier methods** These include condoms (rubber sheaths worn on the penis), female condoms (sheaths worn inside the vagina), and diaphragms and caps that cover the cervix. All are best used with spermicidal cream.
◆ **Mechanical methods** An intrauterine device (IUD) or coil is a piece of plastic or metal inserted into the uterus. It kills or hinders sperm and/or prevents the egg from implanting.
◆ **Hormonal methods** These use female sex hormones to prevent ovulation. The 'morning-after' pill uses a female sex hormone to prevent implantation.
◆ **Surgical methods** Clipping or tying the Fallopian tubes and vasectomy (cutting and tying off sections of the sperm ducts) cut the route for eggs and sperm.
◆ **'Natural' methods** These depend on avoiding sex when the woman is most fertile, around ovulation.

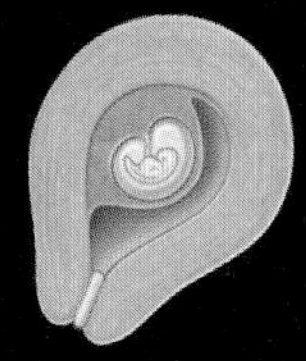

4 WEEKS
The blastocyst grows into an embryo 3 mm (1/8 in) long. The spine, nervous system and muscles are all forming.

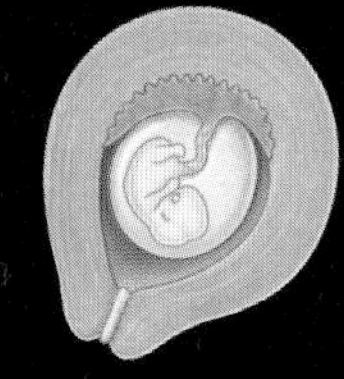

8 WEEKS
The main internal organs have formed and the foetus begins to move.

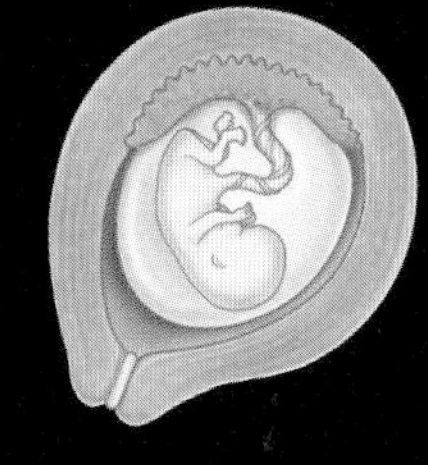

12 WEEKS
The heart is pumping and the pulse appears.

Stages of development

4

The main stages of a baby's development take place in the first 12 weeks, when the mother's body shows few signs of pregnancy. From then, the foetus rapidly increases in size and weight. Its length increases 20-fold from eight weeks to full term (38-40 weeks). Kicking can be felt from about 18-20 weeks. By 22 weeks the baby's internal organs and systems are all functioning more or less fully. This is about the earliest stage at which a premature baby can survive.

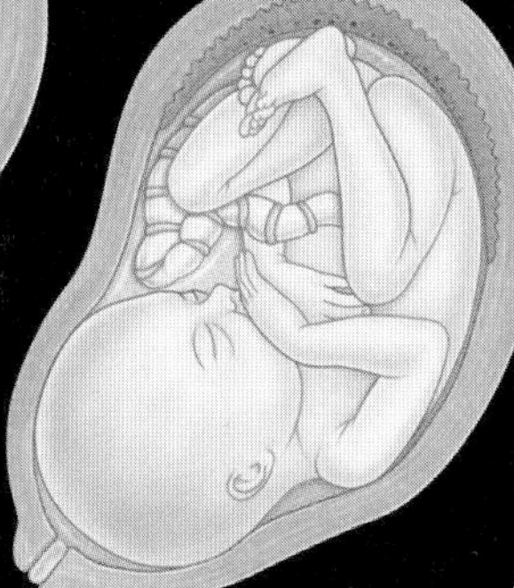

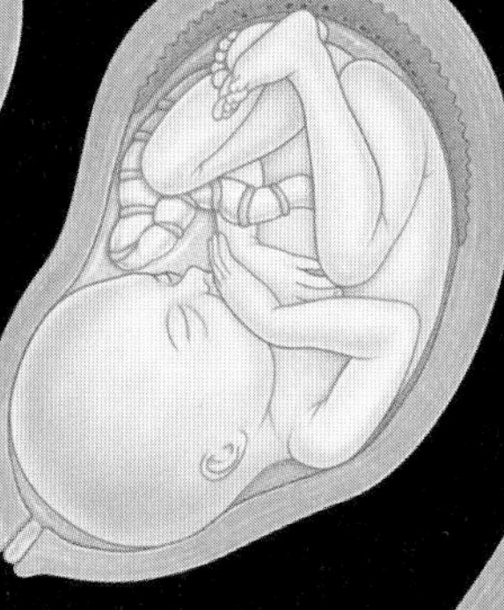

16 WEEKS The lungs are now strong enough to breathe.

22 WEEKS The sex of the foetus can be identified.

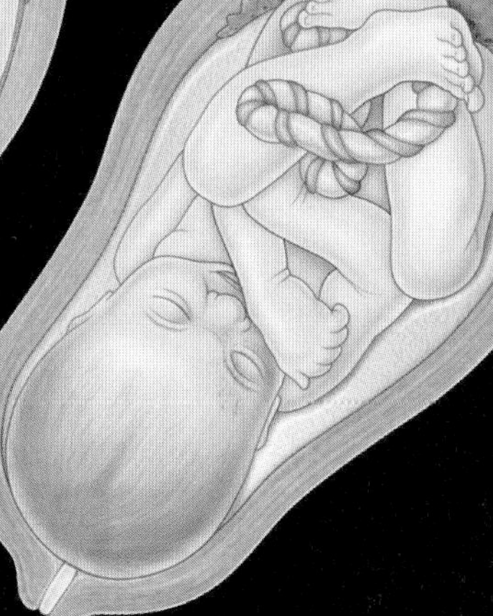

38 WEEKS The foetus, now fully formed, is ready for birth.

Two or more

5

A woman normally releases one egg in each menstrual cycle, but sometimes two or more may be produced. If both or all are fertilised and implant in the uterus, **non-identical (fraternal) twins** or triplets result. Each comes from a separate sperm, so they may or may not be of the same sex.

Less often, a fertilised egg splits into two embryos before or just after implantation in the womb. Since they came from the same sperm and egg, they have the same genes and are **identical (or maternal) same-sex twins**.

Rarely, the embryos only partly divide, giving **conjoined (or 'Siamese') twins**.

Fertility

6

Many problems may affect men and women and prevent a couple conceiving. Common treatments are:

◆ **Artificial insemination** Sperm (from the partner or a donor) is injected directly into the woman's vagina or uterus.

◆ **In-vitro fertilisation (IVF)** Ovulation is induced in the woman and eggs collected from her ovaries. These are fertilised with sperm from her partner or a donor, using a test tube or sterile dish. Several fertilised eggs are then placed in her uterus.

★ 216

X or Y?

The sperm that wins the race to fertilise the egg decides half of the genes the baby will inherit. In particular, it decides the **baby's gender**. Sperm carry either an X or a Y chromosome; if it is a Y the baby will be a boy, if an X a girl. The egg always carries an X chromosome.

Problems in pregnancy

7

In early pregnancy, miscarriages are often due to a genetic defect in the foetus, a lack of the hormone progesterone in the mother, or a weak cervix. Later, mothers are monitored for signs of:

◆ **Pre-eclampsia** High blood pressure, swollen tissues and protein in the urine are signs of this condition, which may cause seizures and death.

◆ **Ectopic pregnancy** The embryo implants in a Fallopian tube insted of the uterus. It bursts the tube as it grows, and the baby is lost.

◆ **Disorders of the placenta** These may cause poor growth, premature birth or the death of the foetus.

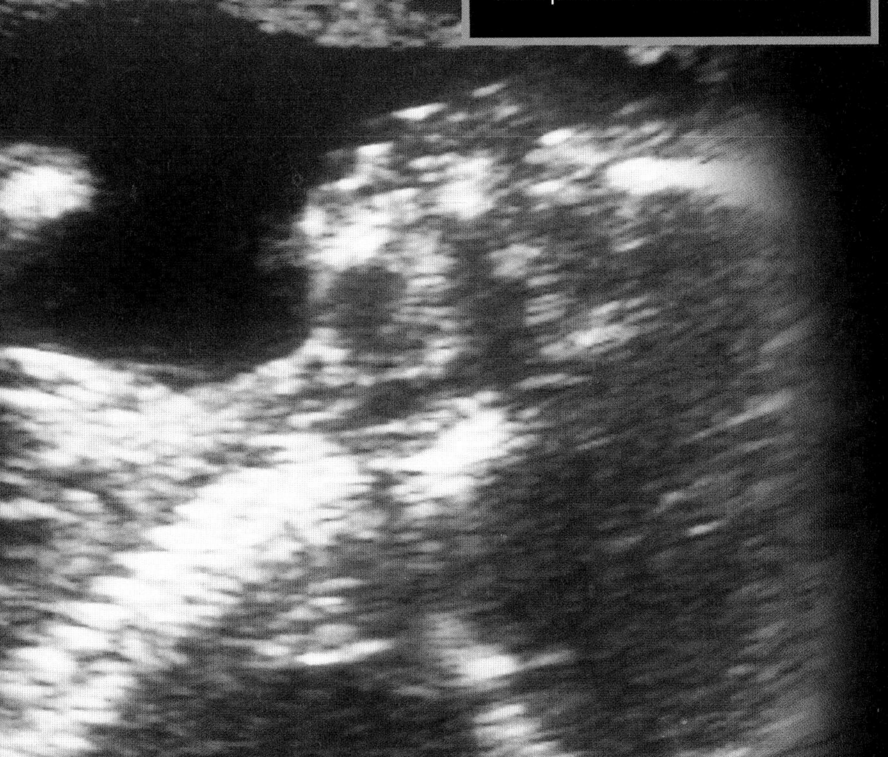

ULTRASOUND SCANS These are carried out during pregnancy to check the development of the foetus and to look for abnormalities.

QUESTION NUMBER

The numbers or star following the answers refer to information boxes on the right.

ANSWERS

47 **5-10 per cent** ❸

198 **True** – 10.2 kg (22 lb 8 oz) born in Italy 1955

278 **The bottom, or the feet** ❸

481 **UK's smallest surviving baby** – 340 g (12 oz)

★ 486 **Julius Caesar**

488 **Queen Victoria** – for birth of Leopold, 1853

571 **Foetal stethoscope** – to monitor baby's heartrate

597 **Obstetrician** – from Latin *obstetrix*, midwife

619 **Causes contractions** ❶

625 **Amniotic fluid** ❷

776 **True** – usually goes within a few days

790 **The placenta and umbilical cord** ❶ ❷

838 **C: 106** ❹

849 **Labour** ❶

895 **Contractions** ❷

965 **True** (after US bacteriologist Robert Guthrie) ❺

989 **Bing Crosby** – his third No. 1 hit of 1938

Childbirth: newborn baby

Core facts ❶

◆ Hormonal changes in a pregnant woman's body trigger childbirth 38-40 weeks after fertilisation. The most powerful hormone is **oxytocin** from the pituitary gland, which causes the walls of the uterus to contract.

◆ **Labour** occurs in three stages, ending with the delivery of the placenta and umbilical cord – known as the **afterbirth**. Most babies are naturally born head-first. A different 'presentation' (birth position) may put the health of the mother and/or baby at risk. Serious complications may lead to a **caesarian**.

◆ Doctors test the **physiological systems** and **hearing** of babies soon after birth.

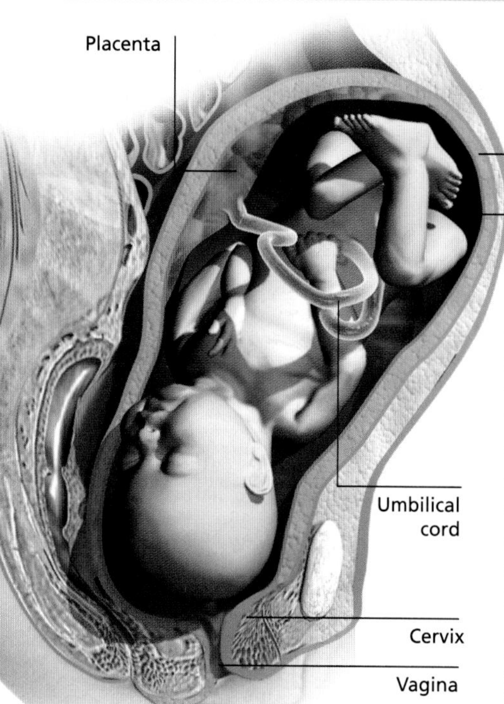

Stages of childbirth ❷

Labour is measured in three stages.

◆ **Stage one** Contractions of the muscular uterine wall push the baby towards the cervix (neck of the womb), which starts to dilate, or widen. This can take 13-14 hours for a first child, less for subsequent children.

◆ **Stage two** When the cervix is dilated to 10 cm (4 in), contractions become stronger and more frequent, and the baby travels down the vagina. The mother can use her abdominal muscles to push the baby out, which takes up to two hours. The umbilical cord is then cut, close to the baby.

◆ **Stage three** The placenta detaches from the uterus. Gentle contractions expel it through the vagina with the rest of the umbilical cord, and help to seal bleeding blood vessels in the uterine lining.

READY FOR BIRTH The head is usually 'engaged' – cradled in the pelvis – when a baby is ready to be born. In stage one, the amniotic sac bursts, releasing the fluid.

Potential problems ❸

Vaginal delivery may need mechanical help with **forceps** (a tong-like instrument) or a **ventouse** (vacuum cup), which holds the baby's head and pulls it along while the mother pushes. Childbirth can tear the skin between the vagina and anus, which is avoided with an **episiotomy** – a deliberate cut which heals more quickly than a tear.

In some situations, vaginal delivery may endanger the health of the mother or baby, and a **caesarian section** may be necessary. **Breech presentation**, in which the baby's buttocks or feet are positioned in the pelvis rather than the head, risks part of the baby entering the cervix before full dilation and becoming stuck. **Prolonged labour** may be a sign that the baby's head is too large to pass through the pelvis; weak contractions which are not strengthened by drugs have the same effect, and surgical intervention may be required. **Poor oxygen supply** can occur if the placenta stops working, or if the umbilical cord tangles around the baby – a complication that can lead to asphyxia.

Between 5 and 10 per cent of babies are **premature** – born before 37 weeks or weighing less than 2.5 kg (5½ lb), before they are fully developed. Prematurity can be caused by hypertension (high blood pressure), kidney or heart disorders or diabetes mellitus in the mother, but often the cause is unknown.

EARLY ARRIVAL A premature baby cannot maintain its body temperature, so must be kept in a warmed incubator and helped to breathe.

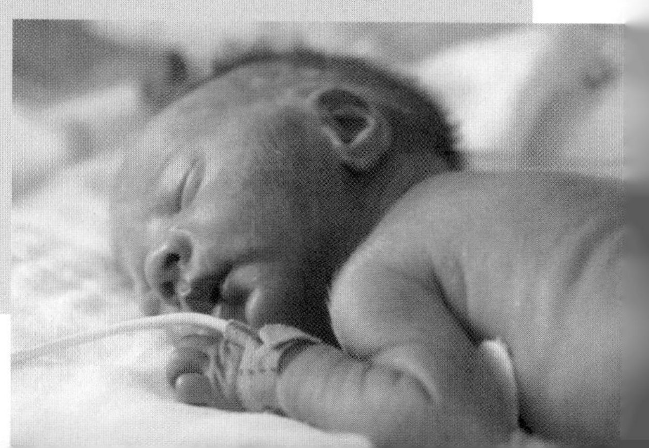

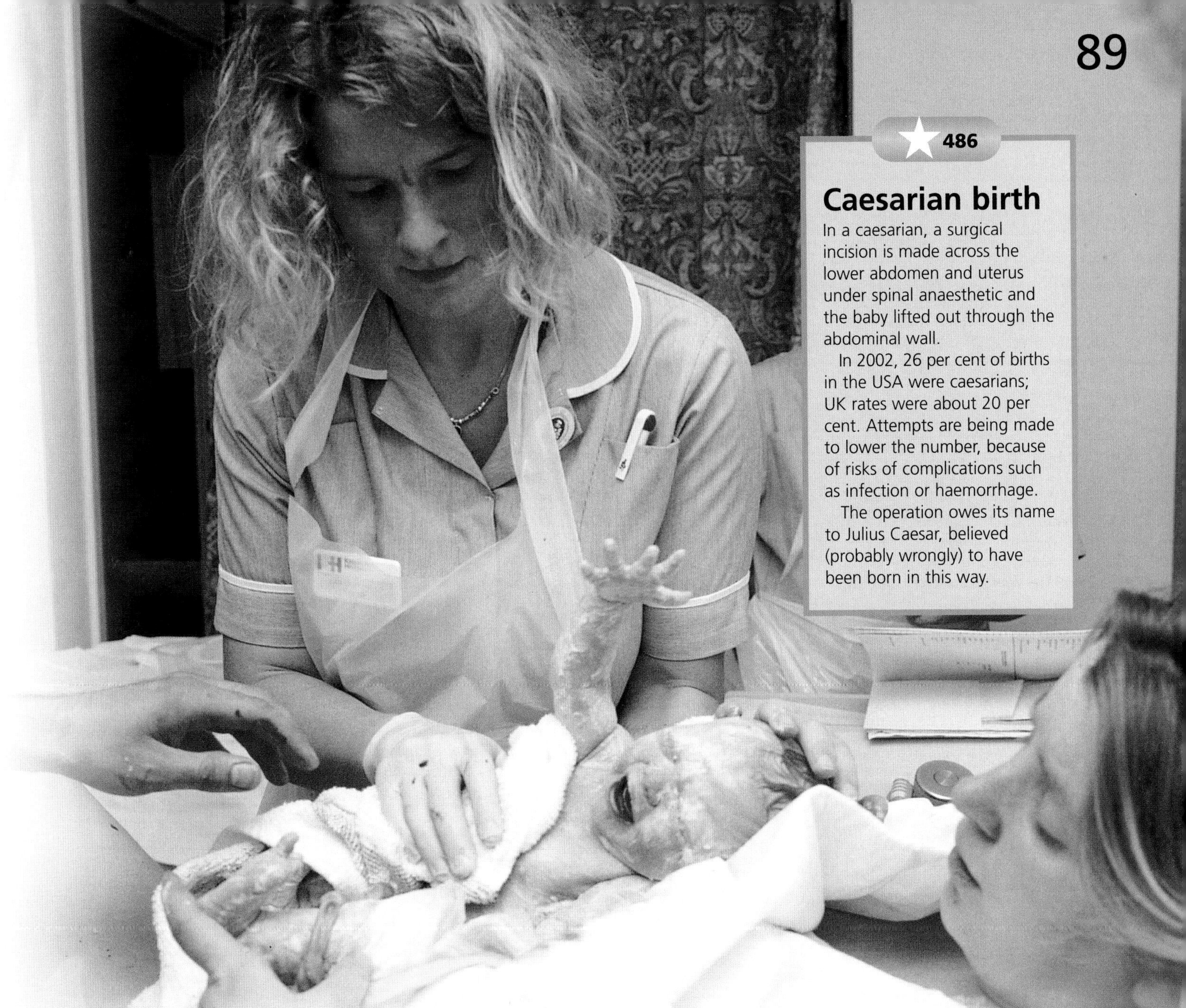

486

Caesarian birth

In a caesarian, a surgical incision is made across the lower abdomen and uterus under spinal anaesthetic and the baby lifted out through the abdominal wall.

In 2002, 26 per cent of births in the USA were caesarians; UK rates were about 20 per cent. Attempts are being made to lower the number, because of risks of complications such as infection or haemorrhage.

The operation owes its name to Julius Caesar, believed (probably wrongly) to have been born in this way.

FIRST BREATH The first inflation of the lungs usually makes a newborn baby cry (above).

4

Birth gender ratio

On average, 106 boys are born for every 100 girls. More boys than girls are stillborn, so the ratio is higher still at conception. On average, men die earlier than women – among people in their thirties the gender ratio is about equal, but ageing can bring about a deterioration in health, and there are more women over 65 than men.

The factors causing the gender ratio have not been explained scientifically. Sperm carrying male 'Y' chromosomes may have an advantage at fertilisation – they may swim faster to reach the egg.

The number of boys conceived is highest in **autumn** and lowest in **spring**, and was especially high during and after the two **World Wars** – facts which scientists have not been able to explain.

5

Tests for newborns

◆ **Apgar score** In 1952, US doctor Virginia Apgar devised a way of assessing a newborn based on heart rate, muscle tone, reflexes, respiration and skin colour. Each factor is scored 0, 1 or 2 (2 being the highest), and the scores added together. The test is performed twice, 1 minute and 5 minutes after birth. An overall score of less than 7 means that a baby needs extra attention and reassessment, until two successive scores of 7 or more have occurred.

◆ **Hearing test** Most hospitals test hearing at two days old by sending a sound into the baby's ear and measuring levels of returned sound.

◆ **Guthrie test** Blood samples collected by pricking the heel of a newborn are tested for a range of disorders such as an underactive thyroid gland and phenylketonuria – an inability to digest a particular protein, which can lead to mental deficiency unless treated through diet.

QUESTION NUMBER

The numbers or star following the answers refer to information boxes on the right.

ANSWERS

48 **3.3 kg/7¼ lb** (accept 3.2-3.4 kg/7-7½ lb) ❷

49 **50 cm/20 in** (accept 45-55 cm/18-22 in) ❷

87 **Measles, mumps and rubella** ❼

115 ***The Catcher in the Rye*** ❻

152 **Eczema** (the only one that is not infectious) ❼

211 **Secondary sexual characteristics** ❺

★ 235 **1940s** (1946)

282 **False** (girls are faster at all these) ❸

349 **Cradle cap** – caused by excess sebum on the scalp

387 **False** (they are taller but lighter) ❹

565 **Lanugo** ❷

566 **Vernix** ❷

604 **Scout** (Jean Louise Finch) ❻

607 **Jamie Bell** – directed by Stephen Daldry

609 **Three times** ❶ ❹

651 **Puberty** ❹ ❺

772 **False** (opposite way round) ❸

962 **True** ❸

972 **Fontanelles** ❷

Birth to maturity

Core facts ❶

◆ When a newborn baby, or **neonate**, leaves the womb, anatomical changes in the heart enable it to adapt to its new environment.

◆ A newborn's bones and organs have different proportions to those of an adult, and these change through childhood as different parts grow at different rates. Development from birth to maturity usually follows a sequence of **developmental milestones**. In the first, body weight triples, then as growth slows, physical and mental skills develop.

◆ Adolescence begins with the physical and emotional upheaval of **puberty**, when the body reaches sexual maturity.

The newborn baby ❷

An average **full-term baby** weighs about 3.3 kg (7¼ lb) at birth, and is 50 cm (20 in) long. Parts of the skeleton are soft cartilage and some organs are disproportionately big.

◆ **Head** A baby's skull bones are separated by fontanelles, or soft membranes. The gaps usually close up by the age of two.

◆ **Bones** At birth, much of the pelvis and limb bones are composed of soft cartilage.

◆ **Heart and lungs** Before birth, the oxygen supply via the blood does not involve the lungs. At birth, partitions develop in the heart to divert blood through the lungs, which start to function with the baby's first breath.

◆ **Thymus** The development of immune cells occurs in the thymus, which increases in size only during the first year of life.

◆ **Liver** The liver is the major site of blood-cell formation before birth, and in a newborn, is large in relation to adult proportions.

◆ **Skin** Premature babies are covered in lanugo (fine hair), which usually disappears at month nine of gestation. At birth, the skin is protected by a greasy substance called vernix.

◆ **Genitals** These are large in a newborn in relation to body size.

Gender and development ❸

Boys and girls differ in average rates of development. Some variations may result from a child's interaction with the environment, including parental attitudes towards the different genders.

Boys tend to cry more at first, and often learn bladder and bowel control later than girls. Growth usually occurs in spurts. Boys tend to be more aggressive and less able to cope with stress, with more behavioural problems. Many boys develop better spatial awareness than girls.

Girls usually develop walking, jumping, hopping and language skills faster than boys, and are more sociable. They grow more quickly, physically and emotionally, and reach puberty sooner.

STAGES OF GROWTH Most children take their first steps (right) by the age of one. At nursery and primary school age, interaction with siblings and peers (above) plays a vital role in social development.

Milestones of development

THE CHANGING BODY Body proportions change dramatically in the growth from infant to adult. The developmental milestones detailed here are based on average sizes and weights – every child is different.

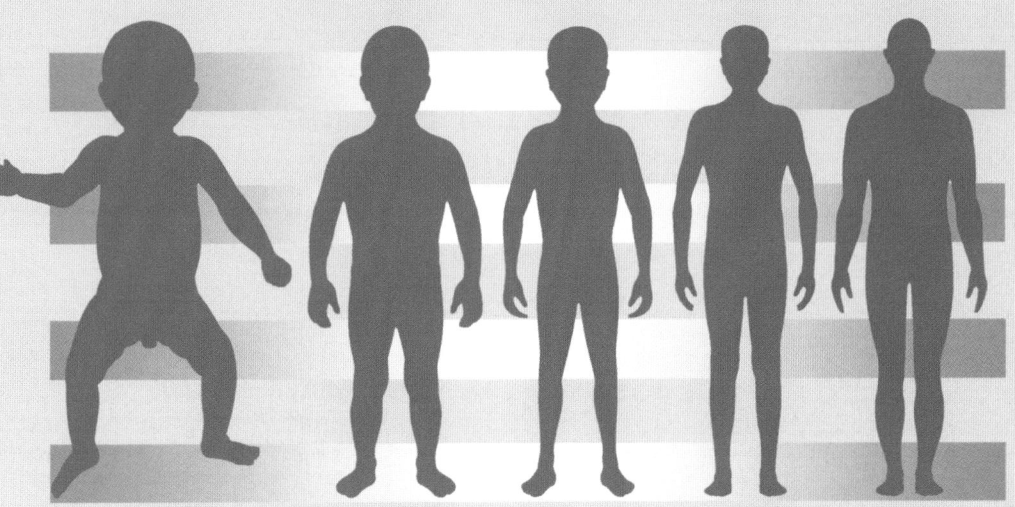

▶ **Birth** *50 cm (20 in); 3.3 kg (7¼ lb)* Grasps; sucks.
▶ **2 months** *58 cm (23 in); 5.5 kg (12 lb)* Squeals; smiles; holds hands together.
▶ **6 months** *68 cm (27 in); 7.8 kg (17 lb)* Reaches out; starts to sit; turns towards voices.
▶ **9 months** *72 cm (28½ in); 9 kg (20 lb)* Crawls or shuffles; grasps; pulls up onto feet.
▶ **c.1 year** *76 cm (30 in); 10 kg (22 lb)* Stands; may walk; forms words.
▶ **2-3 years** *87-96 cm (34-38 in); 12-14.5 kg (26-32 lb)* Throws; kicks; forms sentences; out of nappies.
▶ **3-4 years** *96-102 cm (38-40 in); 14.5-16.5 kg (32-36 lb)* Eats with cutlery; draws people; copies writing; balances; catches.
▶ **4-5 years** *1.02-1.1 m (40-43 in); 16.5-18.5 kg (36-41 lb)* Coordination improves; socialises; reads; writes.
▶ **7 years** *1.22 m (4 ft); 23 kg (3 st 9 lb)* Physical skills present; social skills improve.
▶ **10 years** *1.38 m (4 ft 6 in); 31 kg (4 st 12 lb)* Puberty starts with growth spurt. Girls start periods 2½ years later.
▶ **14 years** *1.58-1.61 m (5 ft 2 in-5 ft 3 in); 48.5-50 kg (7 st 9 lb-7 st 12 lb)* Average boys taller but lighter than girls.
▶ **18 years** *1.64-1.77 m (5 ft 4½ in-5 ft 10 in); 57-69 kg (8 st 13 lb-10 st 12 lb)*. Boys now taller and heavier than girls.

Puberty

Between the ages of 9 and 14, the **pituitary gland** begins to produce hormones which affect the reproductive organs. The ovaries in girls and the testes (testicles) in boys start to manufacture large amounts of sex hormones – **oestrogen** and **progesterone** in girls and androgens such as **testosterone** in boys. These hormones initiate changes in physical characteristics. Girls develop breasts and wider hips. The ovaries mature and start to produce eggs, and menstrual cycles begin. Boys develop stronger muscles, and the voice 'breaks' or deepens as the larynx (voice box) grows. The testes begin producing sperm. Both sexes grow underarm and pubic hair. The obvious bodily changes associated with puberty are known as **secondary sexual characteristics**, and indicate that the body has reached sexual maturity.

TEENAGE CHANGES Some children reach puberty earlier than others – the age also varies between cultures. Since 1840, health improvements have led to earlier maturity.

★ 235

Child care

The best-known childcare authors of the 20th century include:
◆ **Arnold Gesell** *The First Five Years of Life* (1940); *The Child from Five to Ten* (1946)
◆ **Benjamin Spock** *The Common Sense Book of Child Care* (1946)
◆ **Penelope Leach** *Babyhood* (1974); *Your Baby and Child* (1977)

Growing up

Popular 20th-century novels about growing up include:
◆ *The Catcher in the Rye* by J.D. Salinger (1951)
◆ *The Go-Between* by L.P. Hartley (1953)
◆ *Cider with Rosie* by Laurie Lee (1959)
◆ *Goodbye Columbus* by Philip Roth (1959)
◆ *To Kill a Mockingbird* by Harper Lee (1960)
◆ *The Perks of Being a Wallflower* by Stephen Chbosky (1999)

Common disorders

Most children are vaccinated against common diseases such as **measles**, **mumps**, **rubella** (German measles) and **whooping cough**. Other common disorders include **eczema**, an inflammatory skin condition; **asthma**, which causes wheezing and difficulty in breathing; **croup**, in which a swollen larynx causes hoarseness and a barking cough; **glue ear**, a middle ear infection causing pus, pain, fever and temporary deafness; **tonsillitis**, an inflammation of the tonsils; and infestation with *Pediculus capitis*, or **head lice**.

QUESTION NUMBER

The numbers or star following the answers refer to information boxes on the right.

ANSWERS

105 **False** (it is low-density LDL cholesterol that is 'bad') ❸

218 **35 to 40** ❼

219 **FSH** ❷

274 **Baldness** ❼

284 **False** (but the hair thins all over the scalp in women) ❼

314 ***Cosby*** – first shown in 1984

315 ***Iris*** (about Iris Murdoch) ❽

324 **Dementia** ❹

350 **Cancers** ❹

601 **122** ❻

605 ***What Ever Happened to Baby Jane?*** – 1962

606 **Henry Fonda and Katharine Hepburn** ❽

608 **Zsa Zsa Gabor** – born Sari Gabor, in 1917

610 **Geriatrics** ❹

686 **True** ❺

743 **The years** – a hit in 1986

★ **839** **D: Premature ageing**

840 **C: 1 in 5** ❹

917 **Height** ❸

Maturity into old age

Core facts ❶

◆ The peak of physical **health and fitness** usually occurs between the ages of 20 and 30. Many body systems start to decline after 30.
◆ In women, the hormonal changes of the **menopause** bring fertility to an end.
◆ Though a few general milestones of the ageing process can be predicted, many people keep fit and active, mentally and physically, into their 80s or even 90s; others decline earlier. The length and quality of old age is affected by **lifestyle** and **genetic inheritance**.
◆ Changes in the structure and functioning of the body's cells are thought to cause the physical processes of **ageing**.

The menopause ❷

In most women, the ovaries gradually stop responding to the pituitary gland's follicle-stimulating hormone (FSH) between the ages of 45 and 55. **Oestrogen** production decreases, and over a period of 3 to 5 years, known as the **climacteric**, menstruation becomes irregular then stops. Lack of oestrogen causes vaginal dryness and loss of skin tone and bone tissue. Oestrogen inhibits FSH production – without it, FSH causes symptoms such as hot flushes and night sweats. **Hormone-replacement therapy** may remedy some of these problems.

★ 839

Early ageing

Gene mutations can quicken the ageing process. People with **Werner's syndrome** carry a mutation that causes them to age prematurely. By their teens they have the wrinkled skin and grey hair of old age; they succumb to cancers or heart disease, and they die young.

Extending longevity ❸

Many exceptionally old people ascribe their longevity to a habit – a glass of wine a day, for example, or always eating breakfast – but many factors are influential:

Heredity People whose parents and grandparents lived long lives are more likely to do so too.

Weight Obesity and high blood pressure shorten life.

Diet The right quantities of nutritious food and plenty of water maximise lifespan.

Exercise Keeping fit and active, in mind and body, helps elderly people live longer.

Cholesterol levels High levels of beneficial, high-density HDL cholesterol and low levels of 'bad' low-density LDL cholesterol, may help to increase lifespan. Cholesterol levels depend on genetics and diet.

Smoking and drinking Drinking excessive amounts of alcohol and smoking will, on average, shorten the lifespan.

ACTION MAN American George Blair took up barefoot waterskiing at 46. He was still waterskiing in 2003, at the age of 88.

Common disorders

4

As age advances, the immune system becomes less efficient and body repair and recovery are slower, especially over the age of 65. The most serious health risks between the ages of 50 and 70 are cardiovascular diseases such as **arteriosclerosis** and **heart-valve disorders**, respiratory diseases such as chronic **bronchitis** and **emphysema**, and **cancer** (the greatest single health risk for women in this age group).

Over 70, **diabetes**, **pneumonia** and **stroke** are also common. Alzheimer's disease and other forms of **dementia** affect about 1 in 20 people in their 70s and about 1 in 5 over 80. As the body's ability to adapt to temperature-change weakens, **hypothermia** is also a danger.

Why we age

5

Scientists do not know precisely what causes the body to age. Several factors may operate at the **cellular level**. A drop in the body's production of adenosine triphosphate – an energy-producing substance – causes mutations in the DNA (genetic material) of cells, hindering their function. Unstable molecules called **free radicals**, produced by the metabolism, are also thought to damage DNA and cell membranes.

Cells may be 'programmed' to replicate (a tissue-repair mechanism) only a certain number of times before dying. The premature ageing of cloned animals, reproduced from ordinary body cells, supports this idea.

Milestones of ageing

7

▶ **20-35 years** Muscle bulk and strength, bone density, lung function and energy needs reach maximum levels, then slowly decline from about age 30. These are the peak reproductive years.

▶ **35-45 years** Men's testosterone levels decrease. The basal (resting) metabolic rate declines by about 5 per cent per decade. Energy needs drop and exercise becomes more important. Learning ability slows.

▶ **45-60 years** The skin loses elasticity. Many men lose hair from the temples and crown (inherited male-pattern baldness). The eye lenses stiffen – close vision is more difficult. Women experience the menopause; in men, the prostate gland may enlarge.

▶ **60-75 years** Bones and muscles weaken. Hearing and sight may deteriorate. Heart and lung efficiency drops, physical reactions slow, and short-term memory is less efficient. Thickening and hardening of the artery walls raises blood pressure.

▶ **Over 75** The heart and lungs cannot sustain extended activity; stamina decreases. Joints become stiff, and the senses are less acute. By 90, 10 per cent of brain tissue may have been lost.

DEFYING AGE Despite advancing years, the stage performances of Mick Jagger, born 1943, remain energetic.

WEIRD AND WONDERFUL

6

Jeanne-Louise Calment of Arles in southern France lived to the oldest authenticated age. She died in August 1997 at the age of 122 years 5 months. She rode a bicycle at 100, and appeared in the film *Vincent and Me* in 1990.

Getting older

8

Ten films on the theme of ageing:

- *About Schmidt (2002)*
- *Cocoon (1985)*
- *Driving Miss Daisy (1989)*
- *Innocence (2000)*
- *Iris (2001)*
- *On Golden Pond (1981)*
- *Shirley Valentine (1989)*
- *The Straight Story (1999)*
- *That's Life (1986)*
- *Wild Strawberries (1957)*

Exercise

QUESTION NUMBER

The numbers or star following the answers refer to information boxes on the right.

ANSWERS

Question	Answer
25	**True** 5
50	**The person's age** 5
245	**Lactic acid** 2
251	**C: 263 kg (580 lb)** 6
286	**True** (any moderate exercise is aerobic) 2
375	**Jim Fixx** – author of *The Complete Book of Running*
381	**False** (at first it often increases weight) 1
485	**The marathon** (by 43 seconds) 6
487	**Arnold Schwarzenegger** – star of *Terminator* films
616	**Rowing; tennis; aerobics** 3
617	**Jazz dancing; squash; golf** 3
747	**Martha and the Vandellas** – re-issued in 1969
756	**Anaerobic** 2
816	**Florence Griffith-Joyner** ('Flo-Jo') 6
★ 854	**False** 3
948	***Chariots of Fire*** – featuring the music of Vangelis
964	**False** (but it is good in many other ways) 3
981	**Bruce Springsteen** – into charts in 1987

Core facts 1

◆ The human body evolved to be far **more physically active** than most people are today, and deteriorates without exercise.

◆ Moderate exercise maintains health, and can help to **reduce body weight** when combined with a balanced low-fat diet. At first exercise may increase weight by building muscle, which is denser than fat. **Fitness levels** can be gauged by measuring the pulse rate.

◆ Body proportions can be a major factor in determining what type of exercise a person is most suited to. High achievers in a particular sport often share a similar **body shape**, or 'somatotype'.

Aerobic and anaerobic 2

The body's cells use **oxygen** to release energy from **glucose** in an aerobic process that generates energy in the muscles and produces carbon dioxide and water as waste. This is how most runners fuel their activity, but a sprinter's legs need energy faster than the body can provide oxygen. A biochemical process known as anaerobic respiration releases energy without using oxygen; the waste product, **lactic acid**, causes stiff muscles.

WEIRD AND WONDERFUL 4

In 2003, Australian scientists announced the discovery of **gene ACTN3**, which is involved in muscle-building and indicates athletic potential. This raises the possibility of using the gene to spot future sports stars.

The benefits of exercise 3

◆ **Aerobic fitness** improves cardiovascular and lung efficiency, and general stamina.

◆ **Bone strength** brings benefits in later life, when the bones tend to lose density.

◆ **Improved coordination** makes skilled movements and balance easier.

◆ **Joint mobility** helps to retain a range of movement in later life (though over-exercising in youth makes later arthritis more likely).

◆ **Muscular endurance and strength** is achieved through exercise.

◆ **Psychological wellbeing** improves as exercise releases endorphins (natural, feel-good substances) in the brain and promotes relaxation.

Activity	Aerobic fitness	Muscle endurance	Muscle strength	Flexibility	Coordination
Aerobics	4 stars	4 stars	3 stars	3 stars	4 stars
Badminton	4	3	2	3	5
Cycling	4	5	3	3	4
Golf (carrying clubs)	2	3	3	3	4
Dancing (energetic)	4	3	3	5	5
Jogging	4	3	3	1	2
Rowing	4	5	5	2	3
Swimming (laps)	4	3	4	4	5
Tennis (energetic)	3	3	4	4	5
Walking	3	3	4	2	2
Yoga	1	3	2	5	4

PUMPING IRON The load-bearing, anaerobic exercise provided by weightlifting is ideal for building muscle strength and endurance.

Measuring fitness

❺

If the body is fit, the **resting pulse rate** (beats per minute, taken before getting up in the morning), should be in the following range:

	20s	50s +
Men	*60–85*	*68–89*
Women	*72–95*	*76–102*

A figure below the lower level suggests high fitness; above the higher level indicates extreme unfitness. During exercise, the heart rate should stay in the 'training zone' – beating at 60–80 per cent of its maximum capacity. To work out a person's maximum heart rate, deduct their age in years from 220.

854

Warming up

Sudden contraction of cold, unstretched muscles can cause stiffness or injury. The body's metabolism has to adjust to exercise with a gradual increase in **blood flow**, **heart rate** and **muscle temperature**. Gentle 'warm-up' exercises, including stretching muscles to their maximum extension, get the body ready for further exertion. A similar 'cool-down' routine to bring the metabolism to rest helps prevent muscle aches the next day.

Fastest and strongest

❻

There are limits to what the human body can achieve, but these are currently unknown. World records represent the ultimate in physical performance to date. The physical build of women means that the highest female achievers score lower than the best men at the same event. World records as they stood in 2003 included:

Marathon running
42.195 km (26 miles 385 yd)
Male: Paul Tergat (Kenya) – 2 hrs 4 min 55 seconds (20.26 km/h; 12.59 mph); Germany, September 28, 2003.
Female: Paula Radcliffe – 2 hrs 15 mins 25 seconds (18.7 km/h; 11.62 mph); England, April 13, 2003.

Shotput – outdoor
Male: Randy Barnes (USA); 23.12 m (75 ft 10 in); Los Angeles, May 20, 1990.
Female: Natalya Lisovskaya (USSR); 22.63 m (74 ft 3 in); Russia, June 7, 1987.

Sprinting – 100 metres outdoor
Male: Tim Montgomery (USA) – 9.78 seconds (36.81 km/h; 22.87 mph); France, September 14, 2002.
Female: Florence Griffith-Joyner – 10.49 seconds; (34.32 km/h; 21.33 mph) USA, July 16, 1998.

Swimming – 50 metres freestyle
Male: Mark Foster (UK) – 21.13 seconds; France, January 28, 2001.
Female: Therese Alshammar (Sweden) – 23.59 seconds; Greece, March 18, 2000.

Weightlifting ('jerk': 2-stage lift)
Male: Hossein Rezazadeh (Iran) – 263 kg (580 lb); Poland, November 26, 2002.
Female: Tang Gonghong (China) – 167.5 kg (369 lb); South Korea, October 8, 2002.

Natural shape

❼

Athletes with **similar body-shapes** tend to excel at similar events. For example, long-distance runners are usually light and lean, while shot-putters are well-built and muscular.

Few athletes are endomorphic (heavily built with a rounded abdomen and head) – most are mesomorphic (narrow-hipped and muscular) or ectomorphic (thin and angular). A ratio is used to define types of athlete by scoring the three body types – endomorph, mesomorph, ectomorph – out of 7. Long-distance walkers are typically 2–4–4, artistic gymnasts 2–4–3.5 and shot-putters 4–6–2.

MARATHON RUNNER In June 2002, at the age of 28, Paula Radcliffe, received an MBE for her athletic achievements.

THE BIG RACE Since the inaugural race in 1981, more than half a million people have completed the London Marathon.

The numbers or star following the answers refer to information boxes on the right.

QUESTION NUMBER	ANSWERS
22	**True** ❻
★ 283	**True** – neither do other insects
289	**False** (all mammals and most birds do) ★
325	**Narcolepsy** ❽
377	**Sigmund Freud** ❺
461	**Insomnia** ❽
496	**False** ❷
549	**Hypnotic** ❽
558	**Electroencephalogram, or electroencephalograph** ❸
602	***Sleeping Beauty*** – ballet composed by Tchaikovsky
618	**Rapid eye movement** (under the eyelids) ❸ ★
662	**Sigmund Freud** ❺
681	**True** ❶ ❷
682	**False** (people probably dream every night) ❶
748	**A white Christmas** – in the film *Holiday Inn*
899	**Melatonin** ❹
945	***The Big Sleep*** – based on a Raymond Chandler novel
949	***Insomnia*** – directed by Christopher Nolan
956	**C: 11 days** ❻
961	**False** ❶ ❸

Sleep and dreams

Core facts ❶

◆ **Patterns** of sleeping and waking change during childhood and again in old age.
◆ The desire to sleep is triggered by **hormonal mechanisms**. The effects of **deprivation** show that sleep is vital to health.
◆ **Electroencephalography** can be used to measure the activity of the brain during sleep.
◆ Scientists believe that people **dream** each time they sleep, though dreams are not always remembered. Rapid movement of the eyes beneath the eyelids and higher frequency brain waves indicate a period of dreaming.
◆ Evidence of dreaming has been found in **animals** other than humans.

Sleep patterns ❷

Newborn **babies** sleep for brief periods, day and night, for up to about 16 hours a day, usually learning to sleep all night after a few months. The sleep needs of **children** decrease as they grow, from about 14 hours a day at one year to about 9-12 hours from as early as two years old.

Among **adults** sleep needs vary widely, usually between 6 and 9 hours. From about the age of 60, the average decreases to a daily total of 6 hours.

The nature of sleep ❸

◆ Sleep is a state of **unconsciousness** in which the body remains responsive to strong external stimuli such as loud noise or bright light – unlike the deep unconsciousness of a coma or anaesthetic. Voluntary muscles are relaxed and metabolic functions such as **heart rate** and **breathing** slow.

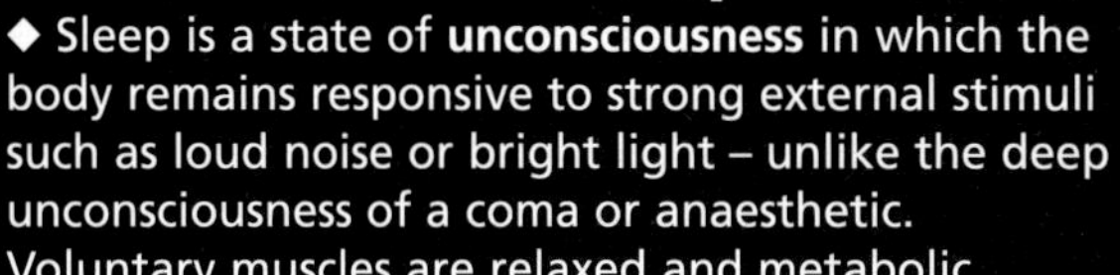

◆ **Brain activity** changes during sleep, but does not always decrease significantly. The measurement of brain waves using electrodes has revealed two main types of sleep – **rapid-eye-movement (REM)** sleep, and **non-REM (NREM)** sleep.
◆ **REM sleep** comes in periods of 5 to 30 minutes a time. Blood flow to the brain is increased, the eyes move beneath the eyelids and the limbs may twitch. People woken from REM sleep usually recall **vivid dreams.**
◆ **Non-REM sleep** is divided into four stages, stage one being light sleep, stage four deep sleep. Metabolic functions and brain activity are reduced, and in stages two and three, the body produces growth hormone – evidence that sleep helps the growth and repair of body tissues.

VISIBLE EVIDENCE Electrical activity in the brain during sleep can be recorded using electroencephalography (EEG), in which electrodes are attached to the scalp.

AWAKE The EEG shows high frequency alpha and beta waves (ten per second).

DRIFTING OFF Relaxing with closed eyes, high frequency beta waves disappear.

FALLING ASLEEP Stage one sleep shows a slight lessening of electrical activity.

GETTING DEEPER In stage three sleep, a lower frequency pattern emerges.

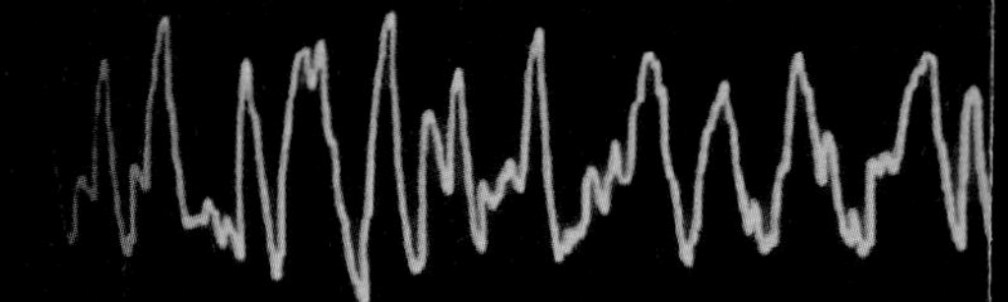

DEEP SLEEP Three or four slow delta waves per second indicate stage four.

REM SLEEP During dreaming, delta waves disappear in a flurry of activity.

DREAMING SLEEP Each time the body moves during REM sleep (below), the electrical activity in the brain undergoes corresponding disturbances.

Body clock

4

The body has a built-in **biological 'clock'** that makes it naturally active by day and sleepy at night. The mechanism is not fully understood, but involves the secretion of a hormone called **melatonin** by the pineal gland in the brain. Melatonin production increases at night, inducing sleepiness, and decreases during the day.

When travelling across **time zones**, the body maintains its usual waking and sleeping rhythms for a while, causing **jet lag** in air travellers. Exposure to strong sunlight helps to 'reset' the body clock by suppressing melatonin release during the daytime.

WEIRD AND WONDERFUL

7

The term **hypnosis**, coined by surgeon James Braid (1795-1860), comes from the Greek word *hupnos*, meaning 'sleep'. It is a state of awareness in which brain wave frequency lowers, causing relaxation rather than sleep.

★ 283 ★

Sleeping dogs

Only **vertebrates** (animals with a backbone) sleep. **Invertebrates** such as insects simply become less active and take rests from time to time.

EEG evidence of changing patterns of brain activity during sleep have not been discovered in **fish** or **amphibians**, but have been found among **reptiles**, **birds** and **mammals** – though reptiles do not seem to enter true REM states. All mammals experience dreaming (REM) sleep and non-REM, slow-wave sleep.

Wakathons

6

Sleep-deprivation studies show that humans need both REM and non-REM sleep. Lack of sleep causes **fatigue** and **irritability**, and consistent deprivation (less than 5 hours' sleep a night) may increase the risk of heart problems. Lack of sleep leads to numerous 'microsleeps' of just a few seconds – only constant activity will prevent full sleep.

Complete sleep deprivation for two days makes mental tasks more difficult. Three days deprivation causes difficulties with thinking, seeing and hearing. Daydreams may be confused with real events, and **hallucinations** may occur. Sleep deprivation for up to 11 days has been recorded, but provoked serious mental symptoms such as **paranoia**.

Sleep disorders

8

◆ **Insomnia** The inability to get to sleep or remain asleep; often a symptom of other problems such as anxiety or discomfort.

◆ **Narcolepsy** Excessive sleepiness, with spells of involuntary sleep several times a day. Can be associated with other disorders such as **cataplexy** (sudden brief paralysis when awake) and **sleep paralysis** (brief paralysis when about to fall asleep).

◆ **Sleep apnoea** The tongue relaxes and blocks the airway until the need to breathe wakes the sufferer; the condition is linked with heart problems.

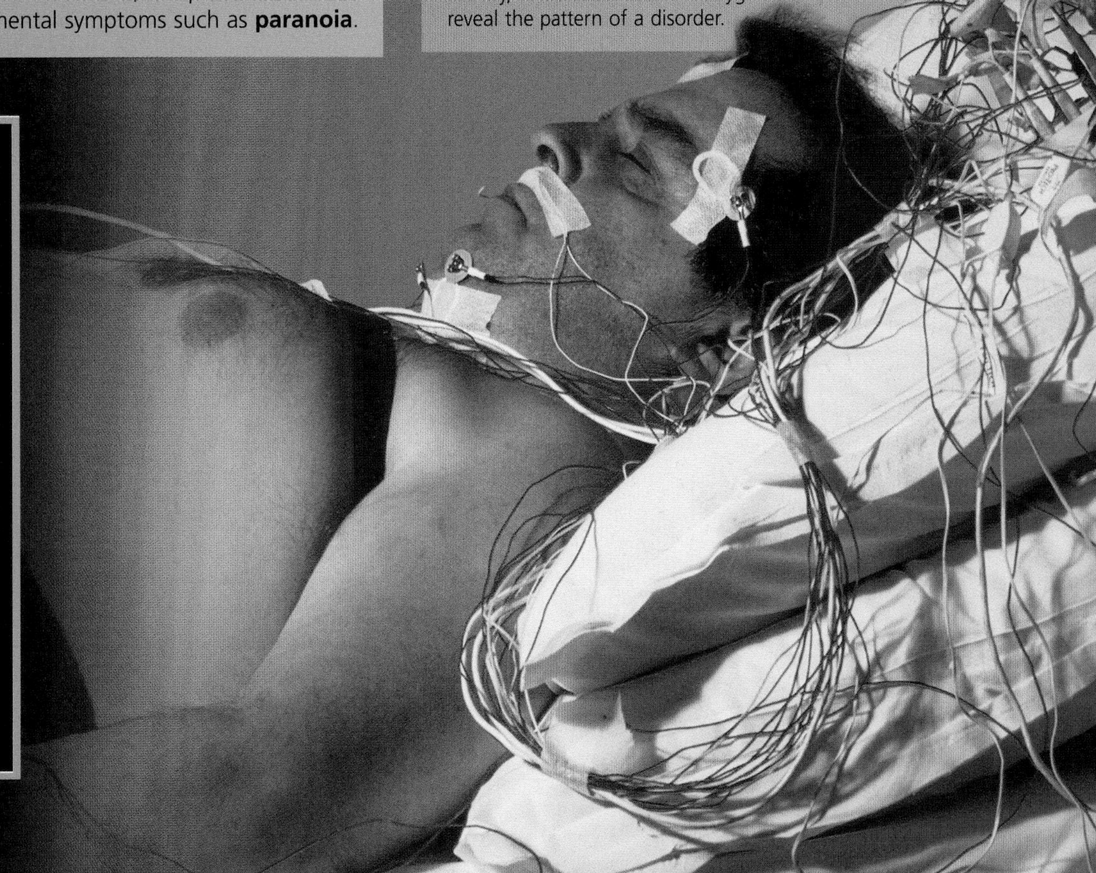

SLEEP INVESTIGATION Measuring brain activity, heartbeat and blood oxygen levels can reveal the pattern of a disorder.

Why dream?

5

In *The Interpretation of Dreams* (1890), Austrian psychoanalyst Sigmund Freud suggested that dreams represent wish-fulfilments, and that objects and situations which appear in dreams are symbols of repressed feelings. Research has shown that areas of the brain activated during dreaming include those which usually respond to emotional and instinctual needs, but many psychologists feel that there is not enough evidence to support Freud's theory, and that dream imagery represents conscious motivations as well as the unconsious mind.

QUESTION NUMBER

The numbers or star following the answers refer to information boxes on the right.

ANSWERS

- **28** **False** – they are two separate disciplines
- **84** **Intelligence quotient** 5
- **373** **René Descartes** (1596-1650)
- **378** **Ivan Pavlov** 4 7
- **509** **Ink-blot patterns** 5
- ★ **563** **Psychology**
- **564** **Cognitive therapy** 2
- **567** **The id** 3
- **568** **Superego** 3
- **659** **Unconscious** 3 4
- **694** **Nicolas Roeg** – also famous for *Don't Look Now*
- **705** **Abnormal psychology** 2
- **706** **Pattern, form or shape** 2
- **757** **Reward and punishment** 3
- **799** **Intelligence tests** 5
- **817** **B.F. Skinner** 3
- **957** **B: Cognitive** 2
- **958** **D: Psychosis** – individual may lose touch with reality

Psychology

Core facts 1

◆ Psychology – the scientific study of human and animal **behaviour** and **mental processes** – emerged in the late 19th century from a background of philosophical and physiological studies reaching back as far as Aristotle.

◆ Psychologists study how **experimental participants** interact with one another and with their **environment**. They try to predict how change affects behaviour, and look for ways of improving individual and group actions.

◆ The major fields of psychology, established during the 20th century, include areas such as **cognitive processing**, **development** and the relationship between **mind and body**.

Major fields of psychology 2

◆ **Behaviourism** A school of psychology based on observing the actions of humans or animals, without reference to mental processes.

◆ **Clinical psychology** The assessment and treatment of psychological disorders.

◆ **Cognitive psychology** The study of mental processes such as decision-making, problem-solving, perception and memory.

◆ **Comparative psychology** The study of the behaviour of different species, comparing humans with other animals.

◆ **Developmental psychology** The study of physical, emotional and environmental changes experienced from birth to death, including child development and gerontology (the study of the elderly).

◆ **Gestalt psychology** A school of psychology which emphasises the overall pattern or form (*gestalt* in German) of behaviour and experience, rather than the individual elements.

◆ **Industrial/organisational psychology** The investigation of organisational behaviour, including personnel practices, work issues and economics. Can involve organisations such as hospitals and civil service agencies, as well as industry.

◆ **Physiological psychology** The study of the relationship between areas of psychology such as behaviour, emotions and perception, and biological functions such as brain and hormone activity.

◆ **Psychoanalysis** A school of psychology founded mainly on the work of Sigmund Freud which emphasises the role of the unconscious mind in shaping human feelings and behaviour.

◆ **Psychopathology** Research into antisocial behaviour and mental disorders such as schizophrenia and depression.

◆ **Social psychology** The study of relationships between individuals and within groups and social institutions.

Psychological glossary 3

◆ **Attention** A tendency to focus on specific features rather than the whole picture.

◆ **Ego** The psychological processes connected with the 'self', or sense of personal identity. Also, Freud's term for the conscious personality, as opposed to the id (unconscious instincts) and superego (the moral conscience).

◆ **Learning** The process by which experience or repetition changes behaviour.

◆ **Motivation** The 'drive' or justification behind a particular behaviour.

◆ **Perception** An awareness or mental interpretation of sensory stimulation.

◆ **Reinforcement** A positive or negative experience that makes a particular response to a stimulus more likely through a process of learning. (Preferred to 'reward' or 'punishment', which project human qualities onto animals.)

◆ **Response** Any action (external or internal) that follows, and is linked to, a stimulus.

◆ **Stimulus** An object or event that elicits a behavioural or physiological response.

RAT LEARNING Using reinforcement techniques, B.F. Skinner showed that rats can learn to press a lever in a box to obtain food.

The pioneers

4

◆ **Ivan Pavlov** (1849-1936) discovered that reflexes can be 'conditioned' to respond to different stimuli and established a basic tenet of learning theory.
◆ **Sigmund Freud** (1856-1939) developed psychoanalytic theory. **Carl Jung** (1875-1961) broke with Freud to establish his theory of the collective unconscious (a cultural inheritance of shared symbols).
◆ **John Watson** (1878-1958) and **B.F. Skinner** (1904-90) founded behaviourism, believing in the observation of physical activity as the only appropriate tool for investigating behaviour and learning.
◆ **Max Wertheimer** (1880-1943) founded Gestaltism with **Kurt Koffka** (1886-1941) and **Wolfgang Köhler** (1887-1967).
◆ **Jean Piaget** (1896-1980) and **Lev Vygotsky** (1896-1934) created influential theories of child development.
◆ **Carl Rogers** (1902-97) founded humanistic psychology with the concept that self-awareness brings well-being.
◆ Cognitive psychology came of age with **Noam Chomsky** (1928-) and his investigations of the mental processes behind language, and through **Herbert Simon** (1916-2001) and **Alan Newell**'s **(1927-92)** computer-modelling of cognitive processes.

THEORIES IN PRACTICE The Swiss psychologist Jean Piaget studied the different stages of children's cognitive development.

TIMESCALE

6

◆ **1875** William James begins teaching psychology at Harvard.
◆ **1879** Wilhelm Wundt founds first experimental psychology laboratory.
◆ **1890s** Sigmund Freud develops psychoanalysis.
◆ **1913** John Watson founds behaviourism.
◆ **1950s-80s** The 'nature-nurture' debate becomes a driving force.
◆ **1956** Dissatisfaction with behaviourism sparks renewed interest in mental processing – the 'cognitive revolution'.
◆ **1960s** Study of visual cortex by David Hubel and Torsten Wiesel revives physiological psychology.
◆ **1970s** CAT and MRI brain scanning boost the study of cognitive skills.

Psychometric testing

5

Psychometric tests measure variables such as intelligence or learning ability. They have practical applications such as screening job applicants, monitoring educational progress and setting parameters for clinical diagnosis. Tests are checked for reliability and validity.
◆ **Intelligence tests** measure skills such as mathematical ability, memory and vocabulary. IQ (intelligence quotient) scores used to be derived by testing for 'mental age', dividing it by actual age and multiplying the figure by 100, but they are now based on statistical IQ averages.
◆ **Aptitude tests** predict future performance based on particular abilities such as mechanical, artistic or interpersonal skills. They are often used in career guidance.
◆ **Personality tests** use questionnaires to measure traits such as extroversion, introversion and neuroticism. The Rorschach (ink-blot) test, which diagnoses mental disorders, asks the individual to interpret ambiguous images, then categorises responses and behaviour exhibited during testing.

BACKGROUND IMAGE A Rorschach ink-blot of the type used to test for emotional stability and intelligence.

WEIRD AND WONDERFUL

7

Pavlov had observed that dogs salivate at the smell of food. He started to ring a bell just before feeding them and after a few times found they salivated at just the sound of the bell. He called this a conditioned (learned) reflex.

The big questions

8

Psychologists are still trying to answer fundamental questions:
◆ What is the physical basis of the mind?
◆ What is consciousness, and can conscious experience be defined?
◆ What is emotion?
◆ What is the physical basis of memory, and why are some things, at some times, easier to remember than others?
◆ Do all people experience the same things in the same way?
◆ Why are some people better than others at some mental tasks?

563

Studying the psyche

The term 'psychology' is derived from two Greek words: *psyche* ('mind' or 'soul') and *logia* ('study'). The Greek philosopher **Aristotle** is often regarded as the originator of psychology. He believed the mind or soul to be separate from the body.

QUESTION NUMBER

The numbers or star following the answers refer to information boxes on the right.

ANSWERS

74 **Small or cramped spaces** 4

122 **Lithium carbonate** 4

130 **Chlorpromazine ('Largactil')** 2 4

141 ***One Flew Over the Cuckoo's Nest*** – 1975

145 **Arachnophobia** 4

147 ***Shine*** – Geoffrey Rush won an Oscar as David Helfgott

149 **Obsessive-complusive disorder** 4

205 **Schizophrenia** 4

253 **A: Argumentative** – may also be chronically late

351 **Mental illness** 2

423 **Psychotherapy** (psychological therapy) 2

468 **Depression** 4

551 **Attention deficit disorder** – affects 1 in 20 children

557 **Electroconvulsive therapy** 2

652 **Phobia** 4

729 **Nerve or nervous** 4

736 **Mental illness** 1 4

750 **Madness** – also famous for the No. 1 *House of Fun*

833 **A: 1 in 3** 1 4

850 **Bedlam** – an asylum for the insane from c.1400

★ 918 **Munchausen**

919 **Manic depression** 4

944 ***Girl, Interrupted*** – also starring Angelina Jolie

Mental health treatment

Core facts 1

◆ Mental illness is a disorder of brain function that affects **behaviour**, **personality**, **thought** or **emotions**. Severe disorders can make rational thought and day-to-day life difficult.

◆ Mental illness affects one in three people to some degree, at some point in life. Treatment includes **drugs** or **psychotherapy**.

◆ Disorders are grouped by their effects, such as **dissociation**, in which a particular mental function is cut off from the rest of the mind.

◆ **Psychiatrists** are doctors who specialise in the prevention, diagnosis and treatment of mental problems; **clinical psychologists** are trained in medical assessment and treatment.

Psychiatric treatment 2

Treatment for mental illnesses is based on **physical intervention** through drugs (somatic therapy) or electroconvulsive therapy, or on **psychotherapy**, also known as 'talking cures'.

◆ **Drugs** Anti-anxiety drugs with calming effects include benzodiazepines such as diazepam, and beta-blockers such as atenolol. Antidepressants such as fluoxitine ('Prozac') change levels of the brain's neurotransmitters (mood-altering chemicals). Antipsychotics such as chlorpromazine ('Largactil') are used to treat serious disorders such as schizophrenia and bipolar disorder (manic depression).

◆ **Electroconvulsive therapy (ECT)** Passing an electric current through part of the brain, causing muscle spasms, can improve symptoms of severe depression after three treatments. ECT is used only in cases with a serious risk of suicide which have not responded to drugs.

◆ **Behaviour therapy** The application of psychological learning theory to re-orientate patients' habitual responses to problems.

◆ **Cognitive therapy** Retraining thought processes to eliminate negative patterns.

◆ **Play therapy** The treatment of disturbed children by encouraging them to tell stories or act out conflicts using toys and dolls.

◆ **Psychodynamic therapy** The patient or client talks – often about whatever comes to mind – with or without the therapist's prompting or questioning. Dreams and childhood memories may be explored. This approach includes psychoanalysis, and often investigates the role of the unconscious in mental problems.

◆ **Rehabilitative therapy** The teaching of new or lost skills or, with severely ill patients, teaching everyday coping skills.

GROUP THERAPY The therapist leads the group session, encouraging participants to help one another by sharing and discussing their experiences.

Tie-breaker 3

Q Which mental illness did Robert Louis Stevenson dramatically portray in *The Strange Case of Dr Jekyll and Mr Hyde*?

A Multiple personality disorder. Dr Jekyll's condition is often mistaken for schizophrenia, which is erroneously known as 'split personality'.

4

Mental disorders

A challenge in mental health is how to distinguish different conditions with similar symptoms. Mental illnesses used to be broadly divided into **neuroses** (mild problems such as persistent anxiety or irrational fears) and **psychoses** (severe disorders preventing normal functioning). Disorders are now more commonly grouped according to their effects.

Condition	Characteristics	Causes	Treatment
ANXIETY DISORDERS AND RELATED CONDITIONS			
Obsessive-compulsive disorder	Obsessive thoughts or fears; compulsion to carry out certain actions repeatedly, such as washing hands or checking locks.	Obsessive-compulsive personality type (inflexible; perfectionist; judgemental); stress; may be hereditary.	Behaviour therapy, often combined with antidepressants. Social support.
Panic disorders	Hyperventilation, dizziness, sweating and chest pains. Includes phobias, in which panic is triggered by specific stimuli such as spiders or enclosed spaces.	Childhood experiences; a particularly stressful event; may be partly hereditary.	Relaxation; counselling; cognitive, group or behaviour therapy; in some cases, antianxiety drugs or antidepressants.
MOOD OR AFFECTIVE DISORDERS			
Bipolar disorder (manic depression)	Spells of depression, alternating with mania (euphoria, high energy and rapid speech; grandiose, unrealistic plans).	Major trauma; may be hereditary.	Antipsychotic drugs during manic stages; antidepressants during depressive stages; lithium as a long-term mood stabiliser; hospitalisation.
Depression	Deep sadness; lack of interest in life; low energy; poor concentration and decision-making ability; early waking; insomnia.	Trauma; physical or mental illness; drugs; hormonal changes such as childbirth.	Antidepressants and/or cognitive therapy. In rare, severe cases electroconvulsive therapy.
PERSONALITY DISORDERS			
Delusional disorder	Delusions, for example of disease, grandeur, guilt or persecution; failure to respond to logical reasoning.	Severe depression; schizophrenia; senile psychosis.	Compulsory hospitalisation, in severe or dangerous cases.
Schizophrenia	Hallucinations, delusions, disorganised speech and behaviour, inappropriate emotional responses, withdrawal.	Possibly heredity or stress. Evidence of link to chemical changes in the brain.	Hospital treatment; antipsychotic drugs; social support. About one in five recover fully after one episode.
DISSOCIATIVE DISORDERS			
Hysterical neurosis (hysteria)	Blindness; deafness; pins and needles; tremors; movement problems; limb paralysis with no discernible cause.	Extreme stress; repressed emotions and fears.	After eliminating all possible physical causes, cognitive or behaviour therapy.
Multiple personality disorder	Abrupt switching between two or more distinct personalities in one person, each unknown to the other or others.	Trauma in early life, such as child abuse.	Psychotherapy to treat the original trauma and bring into awareness/ reintegrate the different personalities.

918

Munchausen

Munchausen's syndrome is the repeated seeking of medical attention for a non-existent or self-inflicted condition. The patient is aware that there is nothing wrong. The cause of the syndrome is unclear – it is thought to be a form of **personality disorder**.

Who's who

5

Alfred Adler (1870-1937) Austrian psychiatrist who developed a theory of personality that emphasised striving to overcome feelings of inferiority; he coined the term 'inferiority complex'.

Jean Charcot (1825-93) French neurologist who studied hysteria and its treatment by hypnosis; a strong influence on Sigmund Freud, who studied with him.

Sigmund Freud (1856-1939) Austrian founder of psychoanalysis and dream interpretation; believed that unconscious motivations caused mental illness.

Erich Fromm (1900-80) German-born psychoanalyst who, with Karen Horney (1885-1952), emphasised social influences on behaviour.

Carl Jung (1875-1961) Swiss psychoanalyst who originated the concepts of 'introversion', 'extroversion' and 'collective' symbolism (images inherited from earlier cultures).

Philippe Pinel (1725-1826) French physician whose psychological approach pioneered humane treatment for mental illness.

ENDURING INFLUENCE Freud believed that the experiences of childhood affected adult mental health.

Milestones of Medicine

QUESTION NUMBER

The numbers or star following the answers refer to information boxes on the right.

ANSWERS

223	**Kos** (or Cos) 2
224	**Italy** (at Salerno) 7
★ 296	**Blood, black and yellow bile, phlegm**
379	**Sir Alexander Fleming** 7
434	**14th century** – 25 million died
436	**1984** (accept 1983-5) 7
489	**Florence Nightingale** 7
561	**Hygiene** – Hygieia was goddess of health
562	**Panacea** – originally from Greek, *panakeia*
592	**The Hippocratic oath** 2
599	**A snake** – coiled round a staff
632	**Epidemiologist** 4 7
633	**Pathologist** 4
635	**Pharmacologist** 4 7
637	**Microbiologist** 4 7
638	**Biochemist** 4
797	**X-rays** 7
881	**Christiaan Barnard** 7
884	**Sir Alexander Fleming** 7
887	**William Harvey** 4 6 7
890	**Louis Pasteur** 4 7

Core facts 1

◆ Today's **medical sciences** are based on five centuries of scientific inquiry into the workings of the body, the causes of disease, and methods of prevention and treatment. Medicine still progresses by research and testing.

◆ Modern medicine grew out of the teachings of **Hippocrates** and **Galen**, from folk traditions, trial and error, insight from observation, and some animal experiments.

◆ Many ancient remedies have a scientific basis. Less reliable ancient theories included the **four humours**, which gave rise to procedures such as blood-letting. This was still in wide use until as late as the 19th century.

Hippocratic oath 2

Hippocrates of Kos (*c*.460-380 BC) is often called the father of medicine. He was one of the first to teach that diseases are not caused by the gods or cured by magic, but have natural causes and so can be studied and treated using natural remedies. He is best known for the physician's oath named after him that includes a pledge to do no harm to a patient.

MEDICAL SAYINGS A detail from a 15th-century French manuscript of Hippocrates' 'Aphorisms'. The Hippocratic Collection consists of some 80 medical scrolls kept in the library at Alexandria.

Galen 3

The Greek-born doctor **Galen of Pergamum** (AD 129-c.216) studied anatomy by dissecting animals and was physician to the gladiators of ancient Rome. He believed in the theory of humours and a threefold circulatory system. After the fall of the Roman empire his ideas were kept alive chiefly by Arab physicians. They were reintroduced into Europe from the 11th century onwards (often through Arabic translations of the original Greek) and became the cornerstone of medieval medical thinking. Some of the European physicians who were inspired by his anatomical researches, such as Andreas Vesalius and William Harvey, would eventually disprove many of Galen's teachings.

Medical sciences 4

◆ **Biochemistry** Study of body chemistry. Pioneers: **Louis Pasteur** (French; 1822-95), **Hans Krebs** (German-British; 1900-81).

◆ **Epidemiology** Study of origins and causes of diseases. Pioneer: **Thomas Sydenham** (English; 1624-89).

◆ **Microbiology** Study of microorganisms such as bacteria and viruses. Pioneers: **Anton van Leeuwenhoek** (Dutch; 1632-1723), **Louis Pasteur** (French, 1822-95), **Robert Koch** (German; 1843-1910).

◆ **Pathology** Study of disease processes and effects on body. Pioneers: **Giovanni Morgagni** (Italian; 1682-1771), **Rudolf Virchow** (German; 1821-1902).

◆ **Pharmacology** Study of effects of drugs and other chemicals on body. Pioneer: **Philippus Aureolus Paracelsus** (Swiss-German; c.1493-1541).

◆ **Physiology** Study of internal body functioning. Pioneers: **William Harvey** (English; 1578-1657), **Johannes Müller** (German; 1801-58).

LOOKING INSIDE An illustration of a dissection from an English manuscript of 1412.

WEIRD AND WONDERFUL 5

Medieval doctors examined blood, sputum and urine for signs of disease. Cloudiness in the top layer of a urine sample was thought to indicate illness rooted in the head, and in the lower layer urinary or genital problems.

Key writings 6

◆ Hippocratic Collection (Greece; c.4th century BC) – it is not known if the texts were really the work of Hippocrates
◆ *Anatomical Procedures*, Galen (2nd century AD)
◆ *Canon of Medicine*, Ibn Sina (known in Europe as Avicenna; 11th century)
◆ *On the Structure of the Human Body*, Andreas Vesalius (1543)
◆ *Great Surgery Book*, Philippus Aureolus Paracelsus (1536)
◆ *Concerning the Motion of the Heart*, William Harvey (1628)
◆ *The Seats and Causes of Diseases*, Giovanni Morgagni (1761)
◆ *Manual of Human Physiology*, Johannes Müller (c.1830)
◆ *Cellular Pathology*, Rudolf Virchov (1858)

296

The humours

The ancient Greeks and Romans taught that health is governed by **four humours** – blood, phlegm, black bile and yellow bile. These corresponded to the **four elements**: earth, air, fire and water. Illnesses were thought to be caused by an imbalance of these humours and elements.

TIMESCALE

▶ ***c*.27th century BC** Imhotep in Egypt is the earliest known doctor – also an architect.

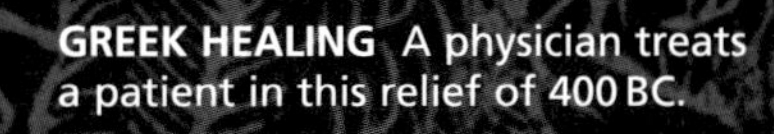

GREEK HEALING A physician treats a patient in this relief of 400 BC.

▶ ***c*.460-380 BC** Hippocrates practises on Greek island of Kos.
▶ **AD 161** Galen practises in Rome.
▶ **10th century** First European medical school at Salerno, Italy.
▶ **11th century** Persian physician Avicenna records details of many drugs.
▶ **1543** Flemish doctor Andreas Vesalius publishes a study of the human body.
▶ **1628** English physician William Harvey discovers blood circulation.
▶ **1670s** Dutch scientist Anton van Leeuwenhoek observes bacteria.
▶ **1726** English chemist Stephen Hales measures blood pressure.
▶ **1796** English physician Edward Jenner performs smallpox vaccination.
▶ **1805** Morphine extracted from opium.
▶ **1815** Stethoscope in use.
▶ **1844-7** Nitrous oxide, ether and chloroform used as anaesthetics.
▶ **1850s-60s** Englishwoman Florence Nightingale establishes modern nursing, and opens nursing school in London.
▶ **1865** English surgeon Joseph Lister pioneers antiseptic surgery.
▶ **1870s-80s** French biologist Louis Pasteur and German doctor Robert Koch develop germ theory of infectious disease.
▶ **1881** Human-to-human blood transfusion by British doctor James Blundell.
▶ **1895** German physicist Wilhelm Roentgen discovers X-rays.
▶ **1899** German company Bayer AG markets aspirin.
▶ **1900s** Vitamins discovered.
▶ **1905** Brain surgery is advanced by US surgeon Harvey Cushing.
▶ **1910** German chemist Paul Ehrlich develops first synthetic antibacterial drug.

ANATOMY LESSON The Dutch artist Rembrandt painted this scene in 1632.

▶ **1915** Existence of viruses confirmed.
▶ ***c*.1915** New Zealand-born surgeon Harold Gillies develops plastic surgery.
▶ **1921** Canadian researchers Frederick Banting and Charles Best isolate a hormone – insulin. Tuberculosis vaccine in use.
▶ **1928** Scottish surgeon Alexander Fleming discovers penicillin.
▶ **1930s** Antibacterial sulpha drugs discovered.
▶ **1935** Drip technique of blood transfusion is used. 7
▶ **1950s** Ultrasound scanning introduced.
▶ **1950** British researchers Richard Doll and Austin Bradford Hill show that smoking is a major cause of lung cancer.
▶ **1953** DNA structure discovered. First practicable heart-lung machine allows open-heart surgery. First kidney transplant.
▶ **1959** Contraceptive pill developed by US physiologist Gregory Pincus.
▶ **1967** First coronary bypass operation. Heart transplants pioneered by South African surgeon Christiaan Barnard.
▶ **1970s** Genetic engineering developed.
▶ **1972** CT (computerised tomography) scanning developed in Britain.
▶ **1973** MRI (magnetic resonance imaging) invented.
▶ **1980s** First successful antiviral drugs produced.
▶ **1982** Insulin made from genetically engineered bacteria.
▶ **1984** Discovery of AIDS virus, HIV.
▶ **1990-1991** First use of gene therapy.
▶ **2000s** Stem cell research promises possible new cures for many diseases.

GENE MAP The complete human genome – a map of human genes – was published in 2003.

Branches of medicine

QUESTION NUMBER

The numbers or star following the answers refer to information boxes on the right.

ANSWERS

82	**General practitioner** ❶ ❷
116	**Tannochbrae** ❹
117	**Holby General** (in Bristol) ❹
142	***The Millionairess*** – based on a play by G.B. Shaw
312	***Dr Kildare*** ❹
317	**Captain Benjamin 'Hawkeye' Pierce** ❹
318	**Mobile Army Surgical Hospital** ❹
319	**4077th** ❹
439	**16th century** ❷
★ 595	**The eyes**
631	**Orthopaedic surgeon** ❺
634	**Urologist** ❺
636	**Haematologist** ❺
639	**Endocrinologist** ❺
640	**Otolaryngologist** ❺
896	**Emergency room (ER)** ❹ ❻

Core facts ❶

◆ Medical knowledge has become so wide-ranging, no one can understand the detail of all areas. Since the 19th century, increasingly specialised fields have arisen, together with a basic **professional division** between general practitioners (GPs) and specialists.

◆ Another core division – between physicians, who 'practise' medicine, and surgeons, who perform surgery – dates back much farther, to the Middle Ages.

◆ Other specialisations are based on **body part**, **type of patient**, or **treatment method**.

Professional divisions ❷

◆ **Physicians** are the original 'doctors'. In Britain, the Royal College of Physicians, their main professional body, was founded in 1518. They include general practitioners and specialists. They may perform minor surgery, but they mainly use non-surgical techniques and drug treatments.

◆ **Surgeons** concentrate on surgery, and may specialise in certain parts of the body or types of disease. Their profession grew out of a medieval 'craft' of barbers, barber-surgeons (who cut hair, pulled teeth and carried out blood-letting) and surgeons. The surgeons split off in 1745 to form the Company of Surgeons – from 1800 the Royal College of Surgeons.

◆ **General practitioners** (GPs) are doctors who deal with routine medical problems and care. In the UK, it is rare for patients to approach specialists – who may work in hospitals or clinics, or private surgeries if fee-charging – directly. Usually, they are referred (introduced) by a GP.

FLORENCE NIGHTINGALE The founder of modern nursing introduced sanitation to military hospitals.

◆ **Nurses** traditionally followed doctors' instructions for patient care, but from the late 20th century they have been increasingly trained to take over routine medical procedures from doctors. During the 20th century, many auxiliary medical professions were formed, such as occupational therapy, prosthetics, physiotherapy, speech and language therapy and radiography.

By type of patient ❸

Specialist	Specialism
Geriatricians	Specialise in treating the ailments of elderly people (geriatrics). (Gerontologists are medical scientists who study such problems.)
Obstetricians	Specialise in treating pregnant women.
Paediatricians	Specialise in the care and treatment of babies and children (pediatrics). May include physicians and surgeons.

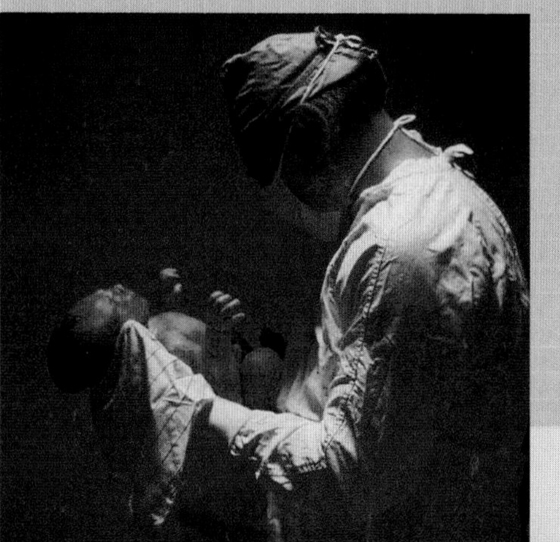

OBSTETRICIAN This is a hospital specialist who deals with childbirth. British hospitals first opened obstetric wards in the 1730s. Today, obstetricians in Britain deal mainly with complications of pregnancy and childbirth.

Screen doctors ❹

Doctor in the House (UK, cinema film; 1954)
Emergency Ward 10 (UK, TV series; 1957-67)
Dr Kildare (USA, TV drama soap; 1961-6)
Dr Finlay's Casebook (UK, TV drama series; 1962-71)
General Hospital (USA, TV soap; 1963-)
M*A*S*H (USA film, 1970, and TV series)
Casualty (UK, TV soap; 1986-)
Peak Practice (UK; TV drama series; 1993-2001)
ER (USA, soap; 1994-)

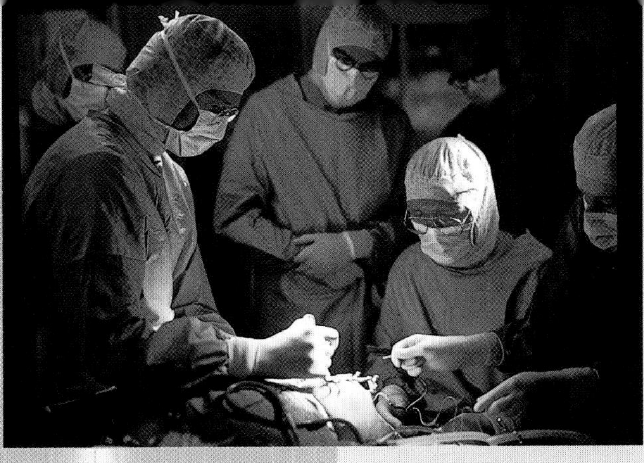

By body part

❺

Speciality	Body part or system
Cardiology	Heart and circulatory system
Colorectal surgery	Lower digestive system
Dental and oral surgery	Teeth and mouth
Dermatology	Skin, nails and hair
Endocrinology	Hormonal system and metabolic disorders
General surgery	Internal organs and other parts of the body not covered by specialist surgeons
Gynaecology	Female reproductive system. Gynaecologists in Britain deal with women's diseases, excluding breast disease. Many also train as obstetricians
Haematology	The blood, including disorders and transfusions
Immunology	The immune system, including allergies, autoimmune diseases and immune deficiency
Internal medicine	The medical equivalent of general surgery, covering areas not the province of specialist physicians
Neurology	Brain, spinal cord and nerves; includes the subspeciality brain surgery
Ophthalmology	The eyes
Orthopaedics	Surgical treatment of bones and joints
Otolaryngology (or ENT)	Ears, nose, and throat
Rheumatology	Medical treatment of joints, ligaments, muscles and tendons
Thoracic or cardiothoracic surgery	Chest, including the heart, lungs, and major blood vessels
Urology	Urinary system (both sexes) and male reproductive system

ORTHOPAEDIC OPERATION
Surgeons operate on an arthritic knee through an endoscope, a tube with a laser beam attached to the end, by passing it through a small hole in the knee joint.

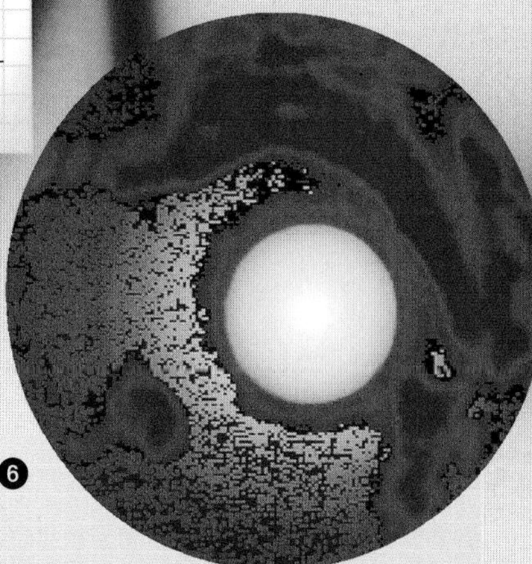

ARTERIAL PLAQUE
This image shows the fatty deposits in a major artery caused by high blood cholesterol. The problem may require cardiac (heart) surgery.

★ 595

Itinerants

Before the 18th century, as well as barber-surgeons there were travelling surgeons, known as itinerants, who specialised in one type of operation. Among these were **occulists**, who treated cataracts; **lithotomists**, who removed bladder stones; and **empirics**, who treated hernias.

By type of illness or treatment

❻

Type of illness	Type of treatment
Accident and emergency (A & E)	Various, as appropriate for injuries, heart attacks, psychiatric emergencies and other traumas
Anaesthesiology	Administration of anaesthetics during surgical operations and other procedures, including life support
Medical genetics	Genetic tests, gene therapy and counselling in cases of illness or potential illness resulting from inherited disorders
Nuclear medicine	Use of radioactive isotopes in diagnosis and treatment of illnesses such as tumours
Oncology	Chemotherapy, radiotherapy and surgery to treat tumours (cancer)
Physiotherapy (physical therapy)	Techniques for overcoming physical disabilities; rehabilitation after injuries, operations, prosthetic limbs and other procedures
Plastic and reconstructive surgery	Repair of skin and rebuilding of tissues after an accident or disease, or for cosmetic reasons
Prosthetics	The use and fitting of artificial body parts, including limbs and joints
Radiology (radiography)	The use of X-rays, ultrasound, CT, PET and MRI scanning and other imaging techniques for diagnosis and treatment – of cancer, for instance
Sports medicine	Physiotherapy, surgery and other techniques to treat injuries caused when practising sports and other physical activities

QUESTION NUMBER

The numbers or star following the answers refer to information boxes on the right.

ANSWERS

QUESTION NUMBER	ANSWERS
75	**Heartbeats** 4
★ 81	**Electrocardiogram, or electrocardiograph**
143	**'Goodness Gracious Me'** 3
232	**1810s** (1816, by René Laënnec) 3 5
255	**A: Auscultation** 3
424	**Palpation** 3
442	**Systolic** 3 5
469	**38°C/100.4°F** (accept 100°F) 4
539	**Reflexes** 3
572	**Stethoscope** 3 5
575	**Opthalmoscope** 5
576	**Ear thermometer** 5
577	**Laryngoscope** – used to observe the larynx
580	**Sphygmomanometer** 3 5
741	**Fever** 4
781	**37°C** 4
883	**Elizabeth Garrett Anderson** – set up London hospital for women (1872)

Diagnosing disease 1

Core facts 1

◆ The first step in treating any disease is an accurate **diagnosis** – deciding what is wrong.
◆ Today, many specialised imaging techniques and tests are available. Even so, observation and a physical **examination** by a trained and experienced doctor are still almost always the starting point.
◆ A **symptom** is an indication of possible illness that the patient can detect. A **sign** is something the doctor observes or measures, but is not apparent to the patient.
◆ Doctors will also need to know something about the **medical history** of the patient and close relatives – and the patient's lifestyle.

Medical history 2

Before examining a patient, a doctor usually asks about:
◆ **Unusual symptoms** or changes in recent weeks.
◆ **Past illnesses, medications and treatments**.
◆ Any **family history** of, for example, heart and circulatory conditions, diabetes and some cancers. Certain tests may be advisable if an illness runs in a patient's family.
◆ **Lifestyle** – alcohol consumption, diet, smoking, stress and foreign travel may increase the risk of certain diseases.

Physical examination 3

The doctor may perform examinations to test or confirm a tentative diagnosis, or may give the patient a general 'physical'. The doctor will observe certain aspects of appearance, including skin, hair, nails, lips and mouth, eyes and joints – in all cases looking for unusual colouring or paleness, swelling, rash and other symptoms and signs. Weight and temperature are also checked. Other routine checks include:

Check	Looking for
Pulse rate	Checking strength and evenness, whether the pulse is 'galloping'.
Breathing rate	Checking whether breathing is easy or laboured, steady or gasping.
Auscultation	Listening to the chest and other sounds through a stethoscope. Heart sounds may reveal valve damage. Wheezing, bubbling and other lung sounds may indicate breathing problems.
Percussion	Tapping with the fingers on the chest or back to check for accumulation of fluid on the lungs.
Blood pressure	Using a sphygmomanometer to check the pressure at the moment when the heart contracts to pump blood (systolic pressure) and when the heart relaxes between beats (lower diastolic pressure).
Palpation	The technique of feeling and pressing with the hands, especially on the abdomen. It may reveal tenderness or unusual swellings. A rectal examination using a gloved finger in the anus checks for growths, an enlarged prostate in men, or lower bowel problems.
Reflexes and other nerve checks	Fully checking nerve function involves a complex series of tests. But a GP may get basic data by checking reflexes – such as the knee-jerk (by tapping just below the kneecap) or the reaction to scratching the sole of the foot. The doctor may also test balance, coordination and strength – for example, by asking the patient to touch the end of the nose, or to follow a moving finger with the eyes.
Other checks	These may include examining the throat and tongue, and checking joints for swelling, pain, and freedom of movement.

THE STETHOSCOPE Doctors use this to listen to the sounds made inside the body by the heart, lungs, blood vessels and bowel.

OBSERVATION AND TOUCH Feeling, looking and listening have always been a vital part of a doctor's physical examination.

Symptoms

❹

One symptom on its own may or may not be important. Doctors look for combinations. Among symptoms that may be significant are deteriorations in faculties such as sight or movement. Others are:

◆ **anxiety or depression**
◆ **coughing; wheezing; shortness of breath**
◆ **diarrhoea; constipation**
◆ **dizziness; fainting**
◆ **fever** – a temperature above normal (37°C/98.6°F) may indicate infection; over 40°C (104°F) is significant.
◆ **forgetfulness, confusion**
◆ **frequent or painful urination; pain during sex**
◆ **lethargy; lack of energy**
◆ **lump or swelling** – whether painful or not
◆ **numbness or tingling**
◆ **pain**, including **headache** – its significance may depend on intensity, location, type, duration, and other symptoms.
◆ **palpitations** (irregular or very fast heartbeats)
◆ **skin symptoms** – itching or a rash (with or without fever) may indicate a skin problem, or an infection or other general condition, such as jaundice.
◆ **trembling or twitching**
◆ **vomiting**, with or without pain, swelling or other symptoms
◆ **weight loss or gain**, with no dietary or lifestyle cause.

CHICKENPOX A rash of flat, red spots is a well-known sign of this childhood disease. It is carried by the virus that causes shingles in adults.

A doctor's instruments

❺

Modern technology gives GPs a bewildering range of instruments to use for diagnosis, but the following have been standard in surgeries for many years:

◆ **Clinical thermometer**
This indicates temperatures between about 35° and 42°C (95° and 108°F). It may be a glass thermometer; an electronic device (mostly used to take the temperature inside the ear); or a plastic strip with patches that change colour with temperature. Although less accurate, the last has the advantage that it sticks to a child's forehead.

EAR PIECE An otoscope or auriscope is a hand-held instrument with an internal light for examining the inside of the ear.

◆ **Ophthalmoscope**
A hand-held viewing device with a built-in light. Its optics and lighting are designed for looking into the eyes.

◆ **Sphygmomanometer**
This measures blood pressure by detecting the air pressure in a cuff wound round the arm to stop blood flow. The cuff is usually placed at the same height as the heart, pumped up, and then deflated. The reading may be shown by the height of a column of mercury in an upright tube.

◆ **Stethoscope** for listening to body sounds, has two earpieces connected by rubber tubing to a body-contact piece. This may have a diaphragm (for high sounds) or a bell-shaped piece (for low sounds) – or both.

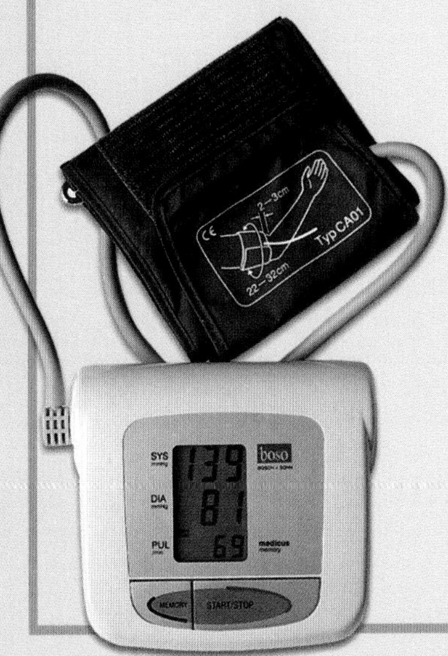

BLOOD PRESSURE MONITOR This gives a digital readout (including pulse rate) on a screen.

★ 81

Electrocardiograph (ECG)

This machine amplifies tiny electrical signals that originate in the heart as beats. The signals are detected by metal electrodes attached to the chest. The machine usually records the pattern of signals as a tracing on paper – an electrocardiogram. The shape and spacing of the 'blips' tells a cardiologist about the patient's heart function.

Tie-breaker

❻

Q What machine is used to measure brainwaves?

A An electroencephalograph (EEG) picks up electrical signals from the brain through sensors placed on the outside of the scalp, and shows them as a tracing on paper.

Diagnosing disease 2

QUESTION NUMBER

The numbers or star following the answers refer to information boxes on the right.

ANSWERS

Question number	Answer
51	Laparoscopy (2) (9)
62	C: PET scan (7) (9)
63	A: CT or CAT scan (3) (4) (9)
238	1910s (2) (9)
300	Infrared or heat ray (5) (9)
443	Blood flow (8) (9)
444	Thermography (5) (9)
★ 445	Endoscopy
447	The heart (8)
476	X-ray (1) (3) (9)
495	True (1) (5) (6) (8)
554	Computerised axial tomography (4) (9)
559	Magnetic resonance imaging (1) (6) (9)
664	Ultrasound scan (1) (8) (9)
714	C: PET scan (7) (9)
846	Ultrasound (1) (8) (9)
875	Contrast medium (3)

Core facts (1)

◆ Many disorders can only be precisely diagnosed by looking at the internal body parts involved. At one time, this meant opening up the body in an **exploratory operation**.

◆ Such surgery may still be performed, but often today by minimally invasive methods, using an **endoscope**, that disturb the inside of the body as little as possible.

◆ Imaging techniques such as **radionuclide scanning** reveal internal parts without an operation, but not necessarily in fine detail. **X-rays**, the oldest method of internal imaging, have been improved and made more versatile.

◆ Newer technologies such as **ultrasound** and **magnetic resonance imaging (MRI)** cut out the risk that exposure to X-rays brings.

Exploratory surgery (2)

When surgery is essential – to remove a tumour, say – a surgeon may perform a minor operation ahead of the main one to make an **assessment**. So before abdominal surgery, a surgeon may perform a **laparoscopy**. This involves cutting a tiny hole in the abdominal wall, inflating the abdomen with carbon dioxide, and examining it through a laparoscope – a specially slender endoscope.

Endoscope

This is a tube that is inserted into a small cut in the skin so that the surgeon can look through it inside the body. The endoscope **beams light** onto the area to be observed, which is then transmitted to the eye by fibre optics or a closed-circuit television camera.

X-rays (3)

These are very **high-energy electromagnetic waves** – like ultra-short-wavelength ultra-violet. They travel through soft body tissues more or less freely, but are absorbed by hard tissues such as bones. The contrast can be picked up on photographic film, or electronically, to give a 'shadow' image of the inside of the body.

CT scans (4)

CT, or computerised (axial) tomography, uses X-rays. It is 'axial' because a patient lies in the centre of a cylinder, with the **X-rays** beamed from many angles, like the spokes of a wheel. Detectors opposite the X-ray source measure the varying strength of the beam as the cylinder rotates. A computer produces a **detailed image** by building up the image 'slices' taken at the points scanned.

SPINE SCAN CT scans are useful for examining the brain and spine. This image shows a slipped disc.

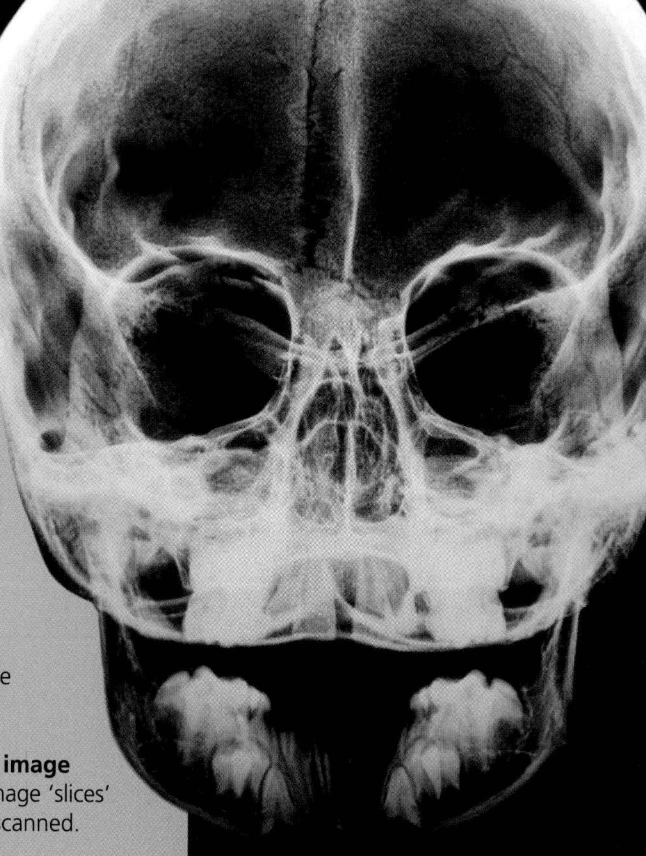

X-RAY SKULL X-ray techinques are relatively simple, but have the disadvantage that X-rays are harmful, especially in large doses.

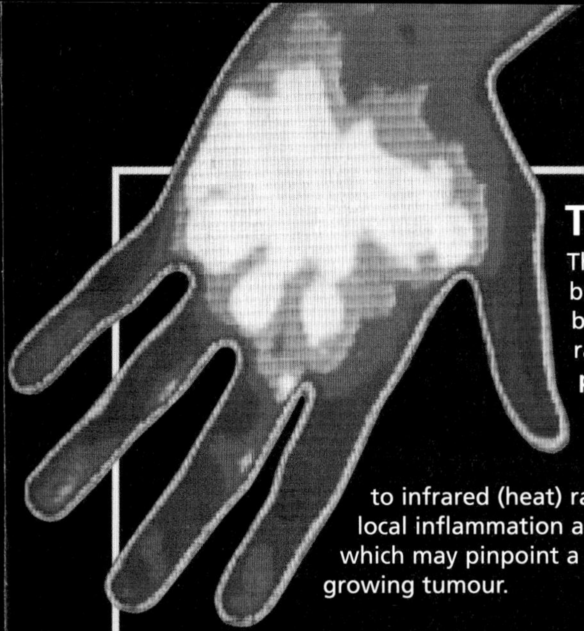

Thermography

5

This is a scanning technique, but it does not involve bombarding the body with radiation. It is a **form of photography** that detects different temperatures on the surface of the skin, using film that is sensitive to infrared (heat) rays. It is useful for detecting local inflammation and rapid cell division, which may pinpoint a growing tumour.

HEAT SENSITIVE Differences in skin temperature appear as colour differences on the image.

Ultrasound

8

Using inaudible, high-pitched ultrasound waves, this technique is very like echo-sounding. It detects and processes the **echoes of ultrasound waves** bounced off internal structures such as tissues. It then produces images on a screen of internal organs or, in a well-known use, of an unborn baby. Ultrasound is especially useful in detecting gallstones, cysts, tumours and blockages such as blood clots. A specialised version, known as echocardiography, is used to look inside the heart.

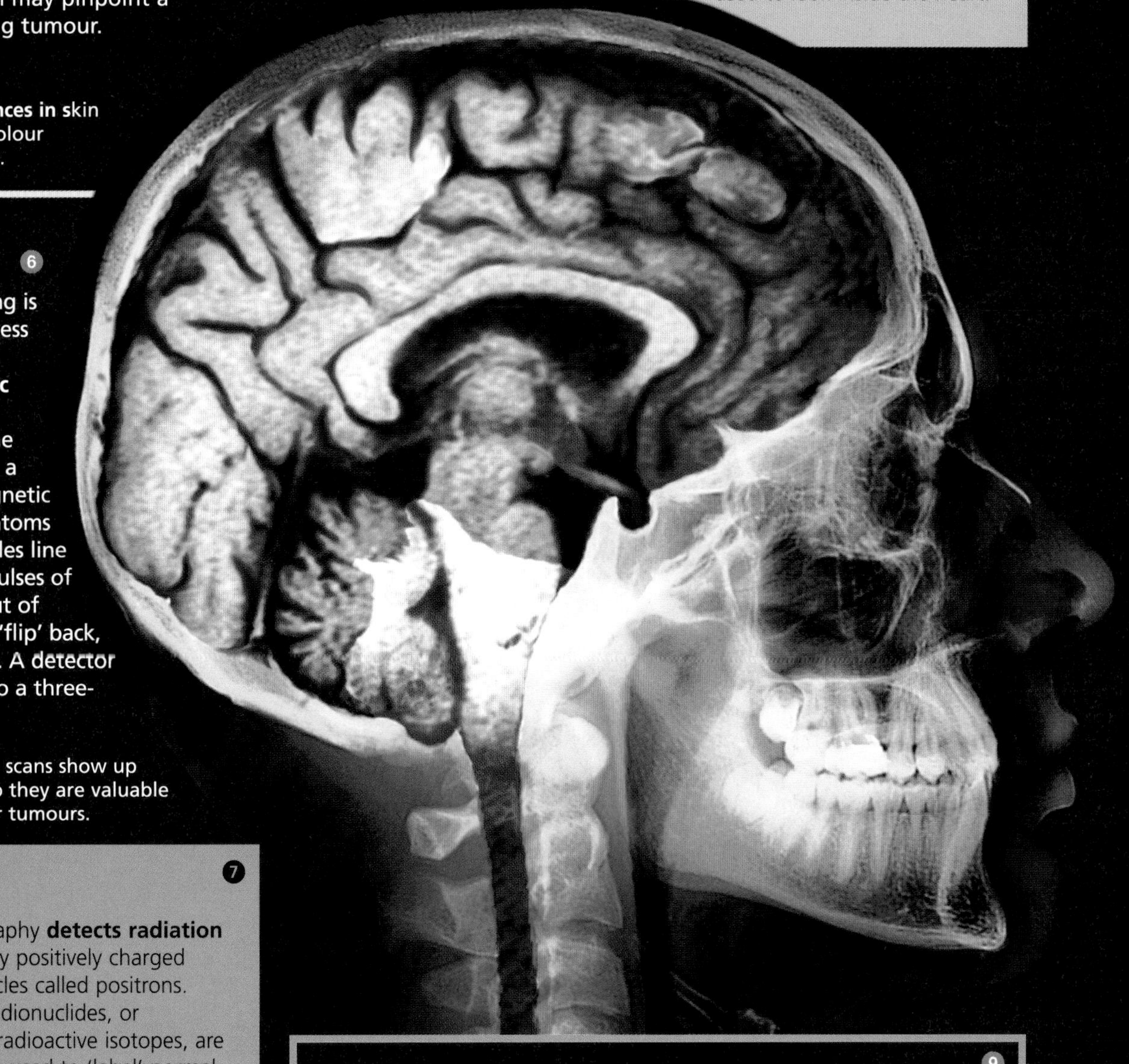

MRI

6

Magnetic resonance imaging is a computer-controlled process that uses the interaction between a **strong magnetic field** and pulses of **high-frequency radio waves**. The patient lies in the centre of a powerful magnet. The magnetic field makes the hydrogen atoms in the body's water molecules line up parallel to each other. Pulses of radio waves knock them out of alignment, and when they 'flip' back, they emit tiny radio signals. A detector magnet processes these into a three-dimensional screen image.

MAGNETIC RESONANCE MRI scans show up hard and soft body tissues, so they are valuable for detecting brain and other tumours.

PET scans

7

Positron **E**mission **T**omography **detects radiation** emitted by positively charged particles called positrons. Radionuclides, or radioactive isotopes, are used to 'label' normal biochemical molecules, and the scan reveals how active various parts of the 'target' organs are. PET is used to examine heart and brain functions.

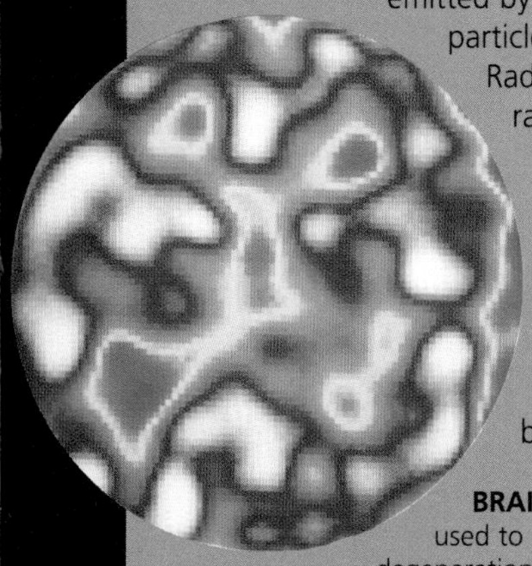

BRAIN SCAN PET scans are used to detect brain tumours and degeneration caused by dementia.

TIMESCALE

9

- ▶ **1895** Wilhelm Roentgen discovers X-rays.
- ▶ **1911** Laparoscopy first used.
- ▶ **c.1950** Ultrasound scanning introduced.
- ▶ **1957** Fibre optics lead to endoscopes with flexible tubes.
- ▶ **1960** Thermography introduced.
- ▶ **1972** CT scanning developed.
- ▶ **1970s** Invention of MRI and PET scanning.
- ▶ **c.1980** MRI scanner in use.
- ▶ **1987** PET scanning developed for medical uses.
- ▶ **1988** Doppler ultrasound scanning introduced.
- ▶ **c.1997** Three-dimensional ultrasound imaging introduced.

Diagnosing disease 3

QUESTION NUMBER

The numbers or star following the answers refer to information boxes on the right.

ANSWERS

52 **Biopsy** ❼

68 **D: Numbers of blood cells** (per mm^3) ❸

172 **Gene or DNA amplification** ❹ ❾

★ 279 **Pregnancy test**

395 **Bacteria** ❸

463 **Diabetes** ❶ ❺

471 **The cervix** ❸

510 **Urine** ❷

540 **Blood type or group** ❻ ❽

582 **Swab** ❷

593 **Screening** ❶ ❽

626 **Antibodies** ❻

680 **Cerebrospinal fluid** (from around the spinal cord) ❷

785 **Blood** ❷

876 **Gene probe** ❽ ❾

Core facts ❶

◆ A vast number of medical **tests** are now available for diagnosis and routine screening.
◆ **Visual tests** are carried out on body fluids and tissues to detect, for example, cell or tissue changes or the presence of pathogens (germs).
◆ Some tests, such as those for **pregnancy** and blood sugar levels, can be carried out at home.
◆ **Screening tests** are performed on people without symptoms to detect conditions such as cervical cancer, for which early diagnosis and treatment can lead to a cure; when there is a risk of an **inherited predisposition** to a condition such as high blood cholesterol; or to detect **abnormal genes** that might be passed on.

Giving samples ❷

Various materials are sampled for different tests:
◆ **Blood** – a drop may be taken from a pin-prick, or a larger sample by syringe from a vein. Sometimes, blood from an artery is needed. The sample may be separated into plasma and cells of different types in a centrifuge.
◆ **Urine** or **faeces** – collected by the patient.
◆ Other **body fluids**, such as saliva, semen, sputum, or pus from a wound or abscess.
◆ Fluids collected using syringes, including **synovial fluid** (from a joint), **cerebrospinal fluid** (from the spinal column) and **amniotic fluid** (from the sac in which a developing baby grows).
◆ **Tissue** and **cell samples**. These may be taken with a swab or scrape from the throat, mouth or vagina, or through a biopsy. Cell samples may also be extracted from body fluids.

★ 279

Pregnancy

A home pregnancy test uses a dipstick to detect a hormone called **human chorionic gonadotrophin** (HCG) in a woman's urine. HCG is produced by a growing placenta. Presence of the hormone is shown by a coloured line; the second line shows that the test is working properly.

Visual tests ❸

Visual examination of fluid or tissue samples is usually done under a microscope. Sometimes the samples are first treated with stains or other materials.
◆ **Smear and thin-film tests** involve examining a thin sample on a glass slide – of blood, for example, or material scraped from a woman's cervix – looking for abnormalities or harmful microorganisms.
◆ A **blood count** estimates the numbers of the different types of blood cell and other particles per cubic millimetre of blood.
◆ Fluids and other samples may be **cultured** – that is, spread on a nutritious jelly and kept warm to allow microorganisms to multiply. These are then stained and examined. The culture may be tested with antibiotics to see which is most effective. Any viruses will be visible only under a high-magnification electron microscope.
◆ **Cell and tissue sample tests** are routine tests that involve treating a sample with preservative. The sample is then cut into very thin slices, which are stained. A doctor or technician examines the slices under a microscope to look for changes that might indicate cancer or inflammation.

SIGN OF CIRRHOSIS
This potentially fatal condition can be detected by examining a small sample of liver tissue, obtained in a biopsy, under a microscope.

WEIRD AND WONDERFUL ❹

In **gene amplification** millions of DNA molecules, all identical copies of the original, are made from one tiny cell sample using polymerase, an enzyme. This enables safe DNA testing of samples from a foetus in the womb.

Chemical tests

Many diseases cause distinctive changes in body chemistry that can be detected by testing a body fluid. Such tests can highlight signs of many problems, including damage to kidney, liver or heart muscle, diabetes, rheumatoid arthritis and bone disorders.

◆ **Chromatography** This uses the different rates at which certain chemicals in a body fluid diffuse through a medium such as filter paper or a column of powdered silica.

◆ **Electrophoresis** Electrically charged molecules, particularly of proteins, are made to diffuse through a liquid or gel under the influence of an electric field. The smaller molecules tend to diffuse faster and farther.

◆ **Dipstick tests** A plastic strip with chemical-impregnated paper patches is dipped into a sample, usually of urine. The patches change colour in the presence of blood, nitrites (from bacteria), proteins, glucose and other material.

◆ **Automatic analyser** This machine uses a light beam and photocell to check for colour changes when samples of urine or blood plasma are combined with various test chemicals.

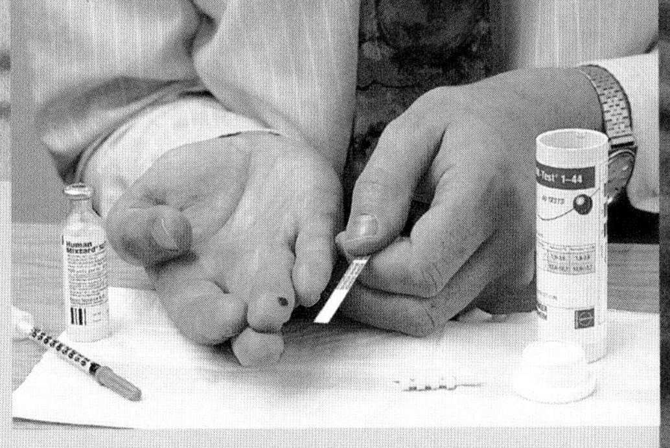

GLUCOSE TESTING Diabetics regularly test their blood glucose levels to check whether and when an insulin injection is needed.

Antigen tests

Foreign substances that stimulate the immune system to produce antibodies are known as antigens, and include viruses, pollen grains and incompatible blood types. Tests can detect the reaction between antibodies and antigens, identifying blood type, allergies or diseases.

◆ **Blood typing** involves mixing a blood sample with antibodies that react against specific blood groups. A reaction is shown by the blood cells clumping together.

ALLERGY TESTING Samples of pollen and other allergens are stuck onto patches and placed on the skin. A skin reaction shows the patient is allergic to a substance.

◆ **Immune function tests** measure the levels of antibodies in a patient's blood. The immune system produces antibodies when it detects the presence of foreign substances, so they can be a sign of immune deficiency disorders and autoimmune diseases.

◆ Other tests detect antigens associated with particular disease organisms such as viruses or bacteria. They also detect the corresponding antibodies in the blood. One type, called an **ELISA (enzyme-linked immuno-sorbent assay) test**, shows a colour change if certain antibodies are present. An **IFA (immuno-fluorescent antibody) test** shows distinctive patterns of fluorescence where antigens and antibodies bind together. Such tests are used to diagnose long-term infections such as HIV, and for checking immunity – for example, to rubella in women planning a pregnancy.

TIMESCALE

▶ **1901-2** US doctor Karl Landsteiner discovers ABO blood groups.

▶ **1903** Chromatography discovered by Russian botanist Michael Tsvett.

▶ **1907** US doctor Reuben Ottenberg first uses cross-matching for blood group compatibility before a transfusion.

▶ **1927** First pregnancy test devised.

▶ **1940** Karl Landsteiner and Alex Weiner discover the Rhesus blood factor, leading to blood-testing of pregnant women.

▶ **1940** Partition chromatography invented by UK chemist Archer Martin. It facilitates analysis of amino acids.

▶ **1952** First use of amniocentesis (amniotic fluid examination) to find foetal abnormalities.

▶ **1983** Huntington's chorea gene mapped, leading to first test for an inherited disease.

▶ **1985** First test to screen blood for the presence of HIV (AIDS) antibodies.

▶ **1993-4** Discovery of genes linked to susceptibility to diseases such as Alzheimer's.

Tie-breaker

Q What is the medical term for surgically removing a small sample of body tissue for further testing?

A Biopsy. A tissue sample from a cyst, for example, may be removed surgically using a biopsy needle, a scalpel, or a specialised tool fed through an endoscope.

Genetic tests

Chromosomal abnormalities can be confirmed – or detected before birth – by examining the chromosomes in sample cells under a microscope. Disorders caused by faulty genes can be detected in several ways:

◆ A **radioactive gene probe** locks on to a known type of abnormal gene. A similar method can be used to detect certain disease organisms by using a gene probe that sticks to part of the organism's unique DNA.

◆ A **DNA test** looks for differences between strands of DNA from the affected person and known normal genes. Abnormal DNA generally breaks into longer or shorter fragments than normal, and this shows up on the DNA 'fingerprint'.

BACKGROUND IMAGE The gene sequence in the DNA of the nematode worm, *Onchocerca volvulus*, a tropical parasite that causes river blindness in humans.

QUESTION NUMBER

The numbers or star following the answers refer to information boxes on the right.

ANSWERS

QUESTION NUMBER	ANSWERS
40	Vectors ❷
326	*Streptococcus*, or streptococcal ❻
396	Pathogens ❶
425	Prions ❹
464	Bacteria and viruses ❶ ❸
550	Antifungal ❸
583	Spread by contact or touch ❶ ❷
584	Fungi ❶ ❸
684	False ❸
738	Lyme disease ❻
739	Scrofula ❻
742	**'Infection'** – album reached No. 1 in USA
★ 793	Malaria
861	Thrush, candidiasis, or candida ❸
862	Botulism ❻
863	Tetanus, or lockjaw ❻
864	Diphtheria ❺ ❻
865	**Amoebic dysentery** (a form of gastroenteritis) ❻
866	**Leprosy/Hansen's disease** – affects skin and nerves
867	Meningitis ❺ ❻
868	**Typhoid** – also known as enteric fever
869	Cholera ❺ ❻
870	Plague ❺

Infectious diseases 1

Core facts ❶

◆ Infection by microorganisms or **germs** is the commonest cause of disease, but not of death.
◆ Many microorganisms are harmless or even beneficial, but others, called **pathogens**, cause disease because they attack body cells directly, causing **contagious diseases**, or produce powerful toxins (poisons).
◆ **Bacteria**, fungi, protozoans and viruses are the main pathogen groups. Larger parasitic creatures such as worms and their larvae can also infest the body and cause disease.
◆ Physical barriers, the immune system and vaccination all **protect** against infection.
◆ **Drug treatment** fights many infections.

Contagion ❷

The main routes by which germs enter the body are:
◆ **Direct contact** – touching an infected person or object. Germs enter the body through cuts, or are passed from hand to nose or mouth, or during sexual intercourse.
◆ **Through the air** – breathing and sneezing expel droplets that may contain germs. These may be breathed in by others.
◆ **Food or drink** contaminated with pathogens that have not been killed by cooking.
◆ **Bites** – many pathogens spend part of their life-cycle in other creatures (called 'vectors'). Mosquitoes, for example, feed on human blood, injecting saliva that may contain pathogens.
◆ **Contaminated instruments and blood** – syringes can inject germs into the bloodstream.

Germs ❸

Viruses are the smallest pathogens. Each group listed below represents, very roughly, a tenfold size increase in length or breadth.
◆ **Viruses** are on the borderline of the living and non-living. They invade living cells to reproduce, and in doing so damage or kill them. A virus consists of little more than a strand of genetic material – DNA or RNA – surrounded by a protein shell.
◆ **Bacteria** are single-celled organisms, classified by their shape. Few are harmful, but those few cause many diseases. They may kill body cells by invading them or releasing toxins that disrupt the cells' biochemical processes.
◆ **Rickettsiae** are small bacteria that cause typhus and other infectious diseases.
◆ **Protozoans** (also called protists) are single-celled organisms that can cause malaria, intestinal infections and some tropical diseases.
◆ **Fungi** – single-celled yeasts and fungi that form filaments called hyphae. They cause lung infections and thrush.
◆ **Parasitic worms** are multicellular animals such as tapeworms. They live and feed inside human intestines, blood vessels and other tissues.

★ 793

Ronald Ross

Indian-born British military doctor Ronald Ross (1857-1932) proved the role of **mosquitoes** in transmitting **malaria** to humans. Ross worked as a doctor in the Indian Army in the 1890s and studied human and bird malaria. He established that malaria parasites have to spend part of their life-cycles in mosquitoes, and was awarded the 1902 Nobel prize for medicine for his discovery.

BACKGROUND IMAGE Anopheles mosquitoes, which carry human malaria, breed in stagnant water in tropical areas, such as this swamp in India. After Ross established that malaria is transmitted by mosquito bites, a campaign began to drain swamps and spray standing water with DDT to kill mosquito larvae.

WEIRD AND WONDERFUL 4

The most basic pathogens may be misshapen natural protein molecules called **prions**. They are thought to cause BSE in cattle and variant CJD (Creutzfeldt-Jakob disease) in humans by making normal prions destroy brain cells.

BACKGROUND IMAGE
Microsporum gyupseum, a fungus that causes ringworm. Inset, right: The parasitic threadworm, most common in moist parts of the tropics.

Treatments 5

Bacteria cause many infectious diseases, but most can be treated with antibiotics. Tetanus and botulism are among the few that cannot, but they can be treated with antitoxins. There are also vaccines against cholera, diphtheria, Q fever, whooping cough, tetanus, TB, some types of meningitis and pneumonia – and even the bubonic plague.

Major diseases caused by bacteria 6

Name	Name of organism	Parts mainly affected	Notes
Botulism	*Clostridium botulinum*	Nervous system	Deadly; from contaminated food
Cholera	*Vibrio cholerae*	Intestines	From contaminated food or water
Cystitis	*Escherichia coli* and others	Bladder	May spread to kidneys (pyelonephritis)
Diphtheria	*Corynebacterium diphtheriae*	Throat; skin	Once common in childhood
Gastroenteritis	*Escherichia coli, Salmonella* species, and others	Stomach and intestines	Food poisoning; from contaminated food or water; also caused by protozoans and viruses
Gonorrhoea	*Neisseria gonorrhoeae*	Genitals; urethra (male)	Sexually transmitted
Helicobacter infection	*Helicobacter pylori*	Stomach	Main cause of stomach ulcers, but often non-pathogenic; may cause stomach cancer
Leptospirosis	*Leptospira* species	General	Caught from animal urine
Listeriosis	*Listeria monocytogenes*	General	From unpasteurised foods
Lyme disease	*Borrelia* species	Skin; joints; others	Caught from tick bites
Meningitis	*Haemophilus influenzae, Neisseria menigitidis,* others	Membranes of brain and spinal cord	Life-threatening, especially if enters bloodstream; also caused by viruses
Non-gonococcal urethritis	*Chlamydia trachomatis,* various others	Genitals; urethra (male)	Sexually transmitted
Pertussis	*Bordetella pertussis*	Lungs; trachea	Also called whooping cough
Pneumonia	Various	Lungs	Also caused by viruses, protozoans and fungi
Q fever	*Coxiella burnetti*	Heart; liver; others	Caught from farm animals
Rocky Mountain spotted fever	*Rickettsia rickettsiae*	Skin; others	A form of typhus; caught from tick bites, not only in Rocky Mountains
Septicaemia	Various	Bloodstream	Blood poisoning
Skin infections	*Staphylococcus aureus*, others	Skin	Include acne, boils, cellulitis, impetigo, stye
Syphilis	*Treponema pallidum*	Genitals; urethra (male)	Sexually transmitted; affects brain in long term
Tetanus	*Clostridium tetani*	Nervous system	Lockjaw; from infection of deep wound
Toxic shock	*Staphylococcus aureus*	General	Associated with tampon use
Tonsillitis	*Streptococcus* species, others	Throat; tonsils	Also caused by viruses
Trachoma	*Chlamydia trachomatis*	Cornea of eye	Common in developing world
Tuberculosis	*Mycobacterium tuberculosis*	Lungs; skin; others	Slow developing; increasing again in UK
Typhus	*Salmonella typhi*	General	From food and water; carriers remain infective

Infectious diseases 2

QUESTION NUMBER

The numbers or star following the answers refer to information boxes on the right.

ANSWERS

Question	Answer
102	True ❹
★ 109	True
125	Aciclovir ❺
128	Zidovudine (AZT) ❺
173	Severe acute respiratory syndrome ❻
285	True ❺
292	Yellow fever ❺
494	True ❸
621	Acquired immune deficiency syndrome ❺
719	D: The nervous system ❺
737	Glandular fever ❺
779	False ❻
791	Influenza epidemics ❹ ❺
860	True ❺
993	Dengue fever ❺
997	Influenza ❺
998	Poliomyelitis ❺

Virus variety ❶

Viruses have many different shapes. The polio virus is **spherical**; others are **rod-shaped** or have heads or tails. Some contain DNA or a DNA-RNA hybrid (if they contain RNA – like the AIDS virus, HIV – they are called **retroviruses**). Many kill the cells they infect; some alter the function of cells so they lose control over cell division and become tumours – a virus causes cervical cancer. Others incorporate part or all of their genetic information into the host cell's DNA, altering it in ways that enable the virus to reproduce.

Structure of a virus ❷

A single virus particle (called a **virion**) has a core of nucleic acid – a chain of genes that carry instructions for reproducing copies of the virus. One or two protective shells of protein (known as **capsids**) surround this nucleus. Some viruses are also enclosed in an envelope or membrane made of protein, carbohydrates and fats. Surface proteins help to bind the virus to its host.

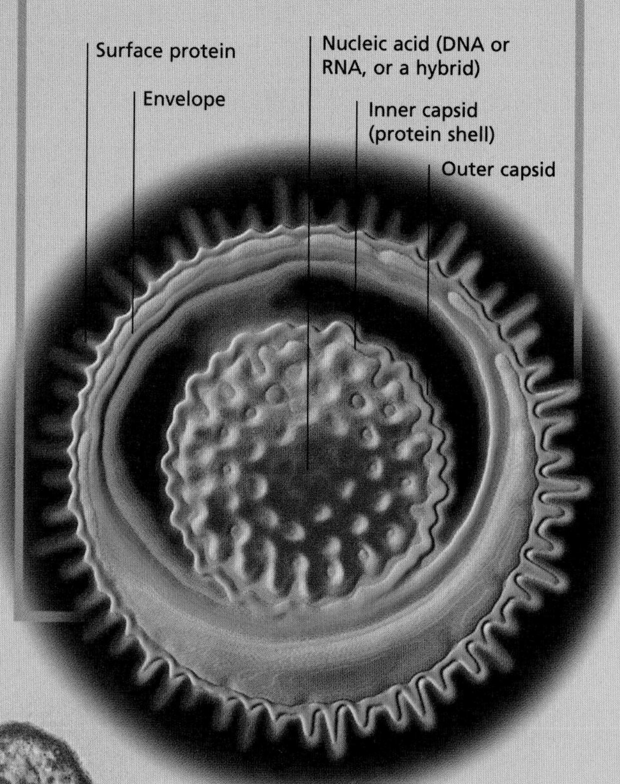

Reproduction ❸

To reproduce, a virus invades a living cell or 'injects' it with its genetic material. The genetic material reproduces itself and **creates new viruses** using the cell's ribosomes (protein-synthesis bodies). These viruses then 'bud' from the cell's surface or burst out of its membrane, ready to invade new cells.

MEASLES MAYHEM The helix-shaped virus buds out from cells it has entered. Once inside a cell, the virus reproduces in large numbers.

Moving target ❹

The HIV virus has evolved in the 20 years since it was discovered. This is called **antigenic drift** – slight changes in the virus's protein shell allow it to side-step the body's defences. The same is true of **influenza**. Every year, a slightly different flu vaccine is developed to fight new strains of the flu virus. More dramatic mutations (genetic changes) in the virus may cause a major epidemic – such as the Spanish flu of 1918 – because no one has immunity and entirely new vaccines would be needed.

Major viral diseases

DEADLY FEVER Bleeding and fever are symptoms of the Ebola virus (below). 5

Name	Name of organism	Parts mainly affected	Notes
AIDS	Human immunodeficiency virus (HIV)	Immune system	Transmitted sexually or by contaminated blood or needles; initial symptoms flu-like
Chickenpox	*Varicella zoster* virus (VZV)	General; skin	Vaccine not generally used in UK
Conjunctivitis	Various	Conjunctiva of eyes	Also non-viral causes such as bacteria and allergy
Dengue fever	Dengue virus	General	Mosquito-borne; up to 100 million cases a year
Haemorrhagic fevers	Hantavirus and various others	General; haemorrhaging	Also Ebola, Lassa (commonest) and Marburg fevers; high death rates, especially for Ebola
Hepatitis	Hepatitis viruses (A to E and others)	Liver	A and E from contaminated food or water; others by blood or sexual contact; also non-viral causes
Herpes	Herpes simplex virus (HSV)	Skin and mucous membranes	Causes cold sores, especially on lips; one type causes genital herpes (sexually transmitted)
Influenza (flu)	Influenza A and B viruses	General; respiratory	Virus (especially A) mutates regularly
Measles	Measles virus	General; skin; nose	Vaccine available alone or as part of MMR (Measles, Mumps, Rubella) vaccine
Mononucleosis (glandular fever)	Epstein-Barr virus (EBV)	General; throat; lymph nodes	Recovery often slow and disease can recur
Mumps	Mumps virus	Salivary glands; testes	Vaccine available alone or as part of MMR
Poliomyelitis	Polio virus	General; brain, spinal cord	Now occurs mainly in developing countries
Rabies	Rabies virus	Nervous system	Transmitted by saliva of infected animal through bites; fatal once symptoms appear
Rubella (German measles)	Rubella virus	Skin; lymph nodes; general	Vaccine available alone or as part of MMR
Warts	Human papilloma virus	Skin; genitals	Genital warts sexually transmitted
Yellow fever	Yellow fever virus	Liver; kidneys; haemorrhaging	Mosquito-borne; mostly in Africa, also Asia and South America; name comes from jaundice

★ 109

Rickettsiae

These are small bacteria but, like viruses, they can reproduce only by invading other living cells. Rickettsiae live as **parasites** in the cells of insects that suck human blood, such as ticks, lice, fleas and mites, and are spread to humans by insect bites. In humans they live in the cells that line capillaries, causing swelling and bleeding, as in epidemic typhus fever, which is passed to humans by lice.

PARASITIC BACTERIUM Unlike viruses, rickettsiae have cell walls, they need oxygen to live, and they can be killed by antibiotics.

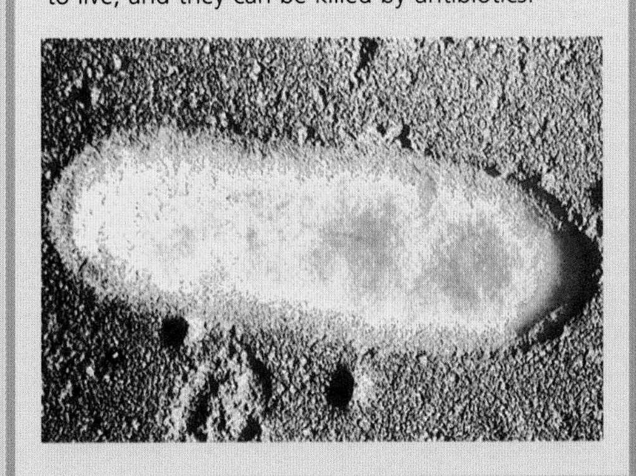

Fighting viruses

6

Many viral diseases can be prevented by **vaccines** because their surface proteins can be recognised by the body's immune system, which then mobilises 'killer' cells to destroy them.

It has been harder to develop drugs that fight viruses without damaging the body cells they invade. Antibiotics have no effect on viruses, but research has enabled **antiviral drugs** to be produced that disrupt the virus's ability to enter body cells or to reproduce inside them.

SARS OUTBREAK The virus responsible for SARS (Severe Acute Respiratory Syndrome) was identified less than a year after the disease was discovered.

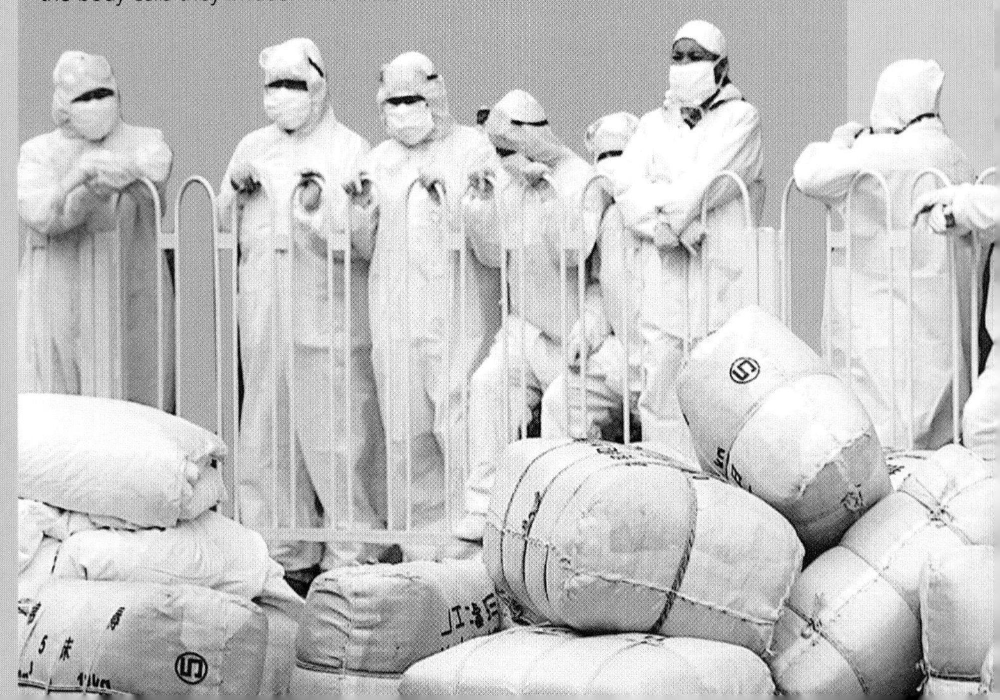

Cancer and heart disease

QUESTION NUMBER

The numbers or star following the answers refer to information boxes on the right.

ANSWERS

- **27** **False** (stoppage is cardiac arrest) 8
- **32** **Crab** – also means 'creeping ulcer'
- **101** **False** (failure = weakening, attack = death of muscle) 8
- **123** **Anastrazole** (inhibits oestrogen production) 6
- **159** **Metastasis** (the others are skin complaints) 1
- **217** **Testicular and prostate** 7
- **327** **Carcinogens** 1
- **343** **Chemotherapy** (often shortened to chemo) 6
- **477** **Lung cancer** 7
- **543** **Cytotoxic drug** (used in chemotherapy) 6
- **547** **Nitrates** (make blood vessels dilate) 8
- **585** **Tumour** 3
- ★ **586** **A balloon**
- **683** **True** 5
- **740** **Secondaries** 2
- **827** **Atheromas** 8
- **878** **Leukaemia and brain or nerve tumour** 7
- **903** **Heart transplant** 8
- **904** **Aspirin** 8
- **920** **Carcinoma** 2
- **940** **True** 5
- **941** ***Erin Brockovitch*** – directed Steven Soderbergh
- **995** **Oedema** – often due to heart or kidney disease

Core facts

◆ The two most common fatal diseases in the UK, USA and other developed countries are **cancer** and **heart disease**.
◆ Cancer causes tumours which can spread through the body in blood and lymphatic fluid – a process known as **metastasis**.
◆ It is not always possible to identify the cause of cancer. Environmental factors thought to provoke its development, such as chemicals or radiation, are known as **carcinogens**.
◆ Heart disease has identifiable causes, such as blocked **arteries** or abnormalities in the **heart muscle**. In many cases, it is preventable or treatable through lifestyle changes or surgery.

Cancer glossary

2

Benign Non-cancerous; describes a tumour that grows but does not spread.
Malignant Cancerous; describes a tumour that can spread to other parts of the body.
Carcinoma Cancer of the epithelial tissue, which lines the organs and forms skin and membranes.
Sarcoma Cancer in connective tissue such as bone, fat or muscle.
Secondaries Cancerous tumours in a different part of the body from the original site.
Oncogenes Genes controlling cell growth that malfunction in cancerous cells.

Tumours

3

Cancer begins with a single mutant (genetically changed) cell which has been either inherited or damaged by carcinogens. The disease becomes more serious as further mutations develop. The main features of cancer are **uncontrolled cell division**, which usually forms a growth known as a tumour, and an **ability to spread** to other parts of the body. Tumours take nutrients from the blood and eventually press against organs and nerves, causing weight loss and pain and interfering with normal body functions.

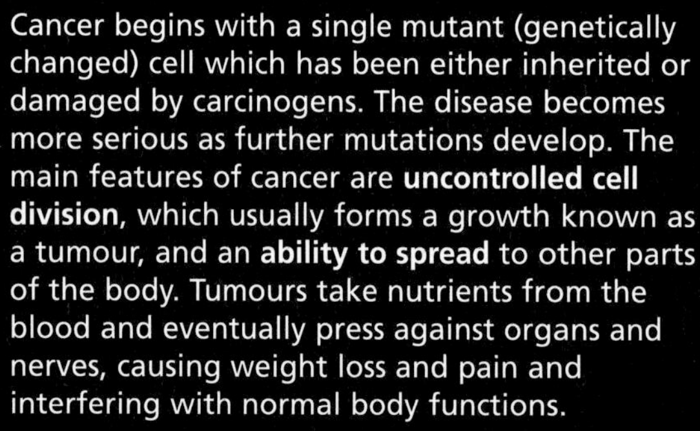

TAKING A CHANCE Smoking is high risk – lung cancer accounts for more than 20 per cent of cancer deaths.

Causes of cancer

4

The cause of most cancers is unknown, though in many cases a combination of genetic and environmental factors are thought to influence its development.
◆ **Food and drink** are believed to play a part in about one-third of cancers. A diet high in animal fats and proteins and low in bulky carbohydrates, fibre and fresh fruit and vegetables is thought to increase the risk, as is a high alcohol intake.
◆ **Industrial and environmental hazards** such as asbestos and nuclear radiation are implicated in relatively few cancers, but can cause a high proportion among those who are regularly exposed to them.
◆ **Tobacco smoke** is a risk factor for about a third of all cancers, not only lung cancer.
◆ **Ultraviolet (UV) rays** in sunlight are known to cause one in ten serious cancers (more in some sunny regions). Skin cancers often develop years after excessive exposure in childhood.
◆ **Viruses** cause about the same number of cancers as UV rays, including most cases of cervical cancer (caused by a type of human papilloma virus, which also causes warts).

How cancer spreads

Tumours grow through **angiogenesis** – the development of new blood vessels which feed it with nutrients. The rate of growth of individual cancers is measured by their **doubling time** (the time needed for the number of cells to double), which ranges from a month to two years. A typical tumour will have doubled up to 30 times, and contain more than a billion cells, by the time it forms a noticeable lump 10-12 mm (about 1/2 in) across. By then, it could have travelled through blood vessels and the lymphatic system into other tissues, causing **secondary** growths.

The spread of cancer is defined by **staging**: Stage 0 cancers are usually self-contained; Stage I and II cancers have spread to nearby tissues; Stage III and IV cancers have spread to distant tissues.

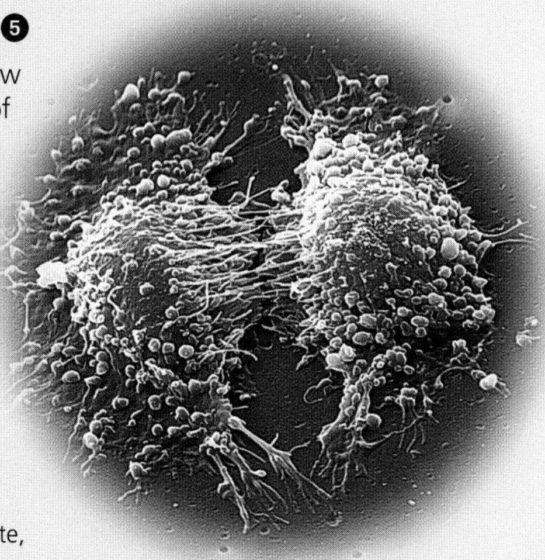

RAPID GROWTH The cytoplasm of tumour cells, like these in the prostate, stretches as they divide and multiply.

Treatment

◆ **Surgery** Cutting away solid cancerous tumours caught at an early stage.
◆ **Radiotherapy** The use of X-rays or other radiation to destroy primary tumours that are difficult to remove surgically, and secondaries.
◆ **Chemotherapy** The administration of drugs that kill cancer cells, often used to treat widespread cancers.
◆ **Hormone therapy** Drug treatment to inhibit hormones that contribute to cancer growth, such as oestrogen and androgen .

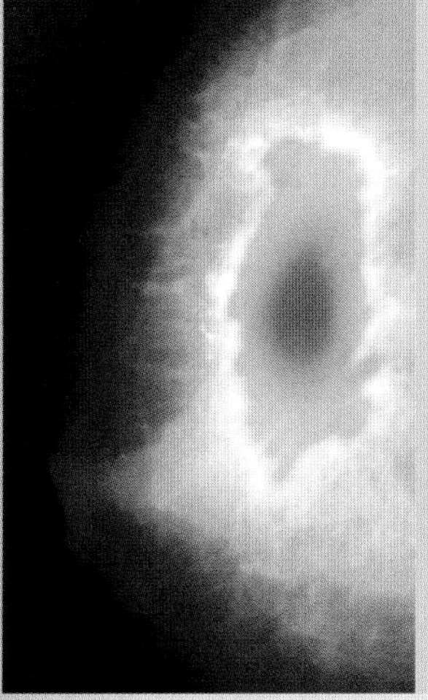

Common cancers

About one in three adults will develop some form of cancer. In children the overall rate is low – about one in 600. There are large international variations in the incidence of common cancers. In the UK and USA, **lung** tumours are the leading cause of cancer deaths. In women, **breast** cancer is the next most common, but fewer than half of those treated die from the disease. **Colorectal** cancer is the next most common, then in men, **prostate** and **bladder** cancer, and in women, cancer of the **uterus**, **ovaries** and **pancreas**.

Other common cancers include **leukaemia** (malignant proliferation of white blood cells), **lymphoma** (cancer of the lymph glands) and **testicular** cancer. These are all sensitive to chemotherapy and usually respond well to treatment.

HARD EVIDENCE A mammogram is a soft-tissue X-ray of the breast, and can detect a tumour (pink).

Heart surgery

586

When the arteries taking blood to the heart are blocked, a **coronary bypass** improves blood supply to the heart by grafting a vein (often taken from the leg) to divert blood around the blockage. In **angioplasty**, a tiny balloon fed in via a catheter is inflated to widen a narrow coronary artery. **Valvuloplasty** is the surgical repair of a defective heart valve. **Lasers** and **suction** are also used to remove artery blockages.

Heart diseases

◆ **Arrhythmias** Disruptions in heart rate and/or rhythm. Tachycardia (an abnormally fast beat), bradycardia (abnormally slow) and fibrillation (uncoordinated contractions) can be caused by a malfunction in the electrical signals passing through the heart muscle, and may be treated with drugs or a pacemaker.
◆ **Atherosclerosis** A build-up of atheromas (fatty deposits) in the lining of arteries. This restricts blood flow, causing angina (chest pain) as the heart muscle is starved of oxygen. It can be prevented or controlled by a diet low in saturated fats, and by drugs called statins.
◆ **Cardiomyopathy** Disease of the heart muscle, such as enlargement of the heart caused by weakening of its muscle (dilated cardiomyopathy) or abnormal thickening in the heart's muscular walls (hypertrophic cardiomyopathy). Serious cardiomyopathy may require a transplant.
◆ **Coronary thrombosis** Blockage of the blood supply in a coronary artery by a clot (thrombus). Lack of blood supply to part of the heart muscle causes that area of the muscle to die – a condition called myocardial infarction, more commonly known as a heart attack.
◆ **Heart failure** Weakening and gradual enlarging of the heart muscle, causing symptoms such as breathlessness and fluid in the lungs. Sometimes treatable with drugs.

KEEPING IN RHYTHM An artificial pacemaker sends electrical impulses along wires to the heart (blue bulge, top left) to maintain a regular beat.

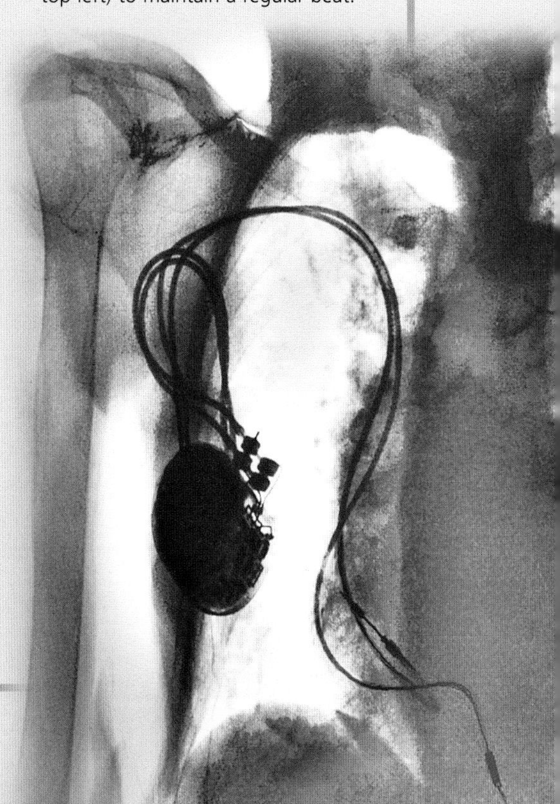

QUESTION NUMBER

The numbers or star following the answers refer to information boxes on the right.

ANSWERS

88 Multiple sclerosis 9

144 *The Singing Detective* 8

254 B: Blocked sweat glands 8

280 Stroke 1 2

288 True ★

328 Cerebral haemorrhage 2

329 Emphysema 3

426 Porphyria 6

483 Stephen Hawking 9

555 Chronic obstructive pulmonary disease 1 3

653 Smoking 3

701 Motor neuron disease 9

735 Parkinson's 10

★ 774 False (alcohol is one of several causes)

858 False (high fat intake in the diet is a major factor) 2

877 Degenerative 1 9

929 Osteoarthritis 7

968 True 7

992 Paralysis (due to a stroke) 2

999 Arthritis 7

Common diseases

Core facts 1

◆ Stroke, known as a **cardiovascular accident** (CVA) by doctors, is the third highest cause of death from disease in the developed world, after heart disease and cancer.

◆ The fourth highest is respiratory disease, including disorders such as chronic bronchitis or emphysema, known collectively as **chronic obstructive pulmonary disease** (COPD).

◆ Other common disorders include **arthritis**, **diabetes**, **skin complaints** and more serious **degenerative diseases** of the nervous system in which symptoms worsen over the years.

◆ Most diseases are caused by a combination of **genetic** and **environmental** factors.

Stroke 2

A stroke occurs when the blood supply to part of the brain is interrupted and brain cells in that area die, causing symptoms such as partial paralysis, numbness, or problems with vision and speech. There are three types of stroke: a blood clot in an artery in the brain is known as a **cerebral thrombosis**; a blood clot which travels from elsewhere in the body and lodges in the brain is a **cerebral embolism**; and the bursting of a brain artery, an event linked with high blood pressure, is a **cerebral haemorrhage**.

Diabetes 5

Diabetes mellitus is an inability to maintain balanced blood-sugar levels. Symptoms include excessive thirst and urination, and tiredness. There are two types. In **type 1 (insulin-dependent) diabetes** the pancreas produces too little insulin; this usually begins in childhood, and treatment involves insulin injections. In **type 2 (non-insulin-dependent) diabetes** the body's cells fail to respond properly to insulin; this usually develops in middle age, and can often be controlled through diet and drugs.

Lung diseases 

Chronic breathing disorders are mainly caused by smoking, and to a lesser extent by air pollution. Signs include a pink, flushed appearance from the effort to breathe, or a bluish complexion from lack of oxygen in the blood.

◆ **Chronic bronchitis** The walls of the airways become thickened, inflamed and congested with phlegm. It becomes difficult to cough, so infection is more likely, which worsens the inflammation.

◆ **Emphysema** The alveoli, or air sacs, at the end of the airways become damaged, stiff and enlarged. The lungs may become distended, resulting in a barrel-chested appearance.

WEIRD AND WONDERFUL 6

Britains's King George III is believed to have suffered from **porphyria**. This inherited condition affects the way porphyrins (metabolic chemicals) are made into haemoglobin, causing skin photosensitivity and mental problems.

Tie-breaker 4

Q A person with diabetes may suffer from excessively low blood sugar levels. What is this condition called?

A Hypoglycaemia, which can be caused by an insulin overdose and relieved by eating sugar. The symptoms include sweating, confusion and slurred speech.

BARREL CHEST Emphysema changes the shape of the lungs and ribs (background image). Air sacs in the lungs (dark areas) enlarge and their walls rupture, and the lung tissue loses its elasticity. This traps air and distends the central parts of the lungs, pushing the ribs apart.

Arthritis

Arthritis is inflammation and stiffness of the joints. Common forms are **osteoarthritis,** in which cartilage between bones wears out and bony outgrowths called osteophytes form, and **rheumatoid arthritis** – inflamed membranes surrounding the joints. Others include **septic arthritis,** caused by an infection; **reactive arthritis,** caused by an immune response to infection; and **ankylosing spondylitis,** in which spinal discs and ligaments become stiff and fibrous.

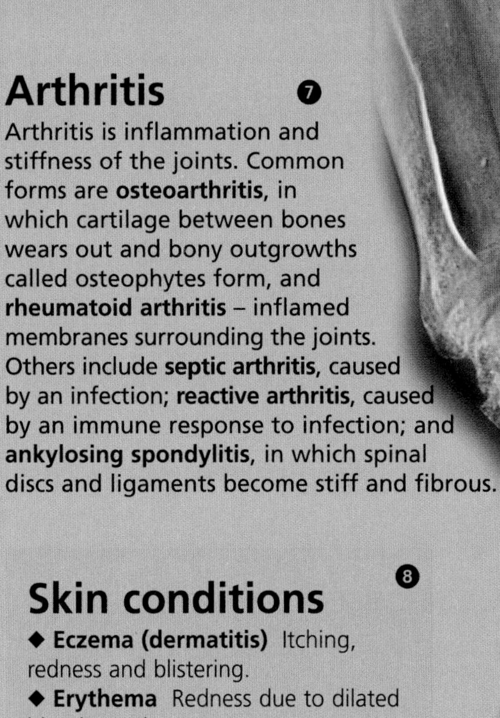

RESTRICTED MOBILITY An X-ray shows the inflamed joints caused by rheumatoid arthritis, an autoimmune disorder in which antibodies made by the immune system attack body tissues.

Skin conditions

- **Eczema (dermatitis)** Itching, redness and blistering.
- **Erythema** Redness due to dilated blood vessels.
- **Impetigo** Itchy rash and blisters caused by bacterial infection.
- **Lichen planus** An itchy rash of small, shiny, flattened spots.
- **Prickly heat** A rash of small, itchy spots caused in hot conditions by blocked sweat glands.
- **Psoriasis** Red areas of skin, covered with dry, silvery scales.
- **Rosacea** Reddening and pimples on the face.
- **Vitiligo** Loss of pigment, in patches; associated with autoimmune diseases.

SIGN OF PSORIASIS The thickened, scaly patches often clear up with exposure to sun.

774

Gout

Caused by raised blood-levels of **uric acid** (a substance in urine), gout is an **arthritic disorder** in which deposits build up in the joints and result in swelling, commonly at the base of the big toe. Gout can be caused by excessive alcohol and rich food, but is also a hereditary condition.

Degenerative diseases

Degenerative disorders are characterised by steady deterioration of nerve tissue and body organs.

- **Alzheimer's disease** Deterioration of brain tissue in middle or old age. Symptoms include memory loss and confusion, and eventually immobility and death. It affects up to 30 per cent of people over the age of 85.
- **Huntington's chorea** Degeneration of nerves deep in the brain that control movement; a rare genetic disorder appearing between the ages of 35 and 50. Causes involuntary muscle movement and dementia.
- **Motor neuron disease** Degeneration of nerves in the brain and spinal cord that control movement, causing muscle wasting. Can be fatal within five years.
- **Multiple sclerosis (MS)** Degeneration of the myelin sheaths that insulate nerves. Affects the conduction of nerve signals and leads to stiffness, tremors and visual disturbances, and eventually to paralysis. Possibly an autoimmune disorder with hereditary factors.

Parkinsonism

Parkinson's disease is a degenerative disorder affecting the **basal ganglia,** an area at the base of the brain involved in maintaining smooth **movements.** It is thought to be caused by a deficiency in a brain chemical called **dopamine,** and results in jerky movements and tremor (shaking) in one or more limbs. Drug treatment can reduce symptoms in the early stages, and drug therapy for the later stages is under investigation.

LOCKED INSIDE Former world-champion boxer Muhammed Ali now displays the rigid expression typical of Parkinsonism.

Traditional medicine: West

QUESTION NUMBER

The numbers or star following the answers refer to information boxes on the right.

ANSWERS

1 **Apothecary** 6

54 **Bleeding or bloodletting** 2

77 **Vinegar and brown paper** 5

78 **Bath** – has the only thermal springs in England

225 **Chelsea Physic Garden** 6

226 **The sauna** – thought to help the circulation

271 **Brimstone and treacle** 5

272 **Sulphur** 5

429 **Porridge** 5

581 **Garlic** (used by herbalists in many remedies) 6

596 **Cupping** 2

598 **To heal cuts** 5

744 **Scaffold** 3

★ 795 **Nicholas Culpeper** 6

879 **Plethora** (from Greek for fullness) 2

973 **Christian Science** (Church of Christ, Scientist) 4

Core facts 1

◆ The ancient female practices of nurturing children and cultivating plants made women skilled in botanical remedies. In the 15th century, however, the Church condemned as witches the **wise women** who dispensed plant cures. Later, male physicians established **herbalism** as an accepted medical practice.

◆ Some of the earliest medical treatments in the West involved **bloodletting** – a practice which could seriously weaken the patient.

◆ **Folk remedies** remained in use as medicine became more sophisticated. The belief that **faith** in a divine power can cure illness has existed for thousands of years.

Bloodletting treatments 2

The ancient Greeks believed that **bad blood** caused disease, and that the best cure was to remove blood from the patient using **leeches** or a **lancet** (knife). Physicians 'let' blood from arteries (arteriotomy) or veins (venesection) by **derivative bleeding** – on the same side of the body as the site of the illness – or by **revulsive bleeding**, from the opposite side. The practice caused weakness and invited infection, but endured until the 19th century.

A less drastic remedy, **dry cupping**, involved applying heated cups to the skin which created a vacuum as they cooled, drawing blood to the surface. In **wet cupping**, blood would be drawn by setting a cup over a small cut in the skin. Dry cupping may have been effective for some problems – it is still used today to ease breathing difficulties and soothe rheumatic pain.

KILL OR CURE? A 15th-century European manuscript records the treatment of the day.

WEIRD AND WONDERFUL 3

The song 'Lily the Pink' is about Lydia Pinkham of Massachusetts, who in 1873 launched a Vegetable Compound to remedy 'all female weaknesses'. It made a fortune – probably because of its high alcohol content.

Faith healing 4

The 'touch doctor' **Valentine Greatrakes** (1629-83) reputedly healed deafness, blindness and cancer by laying his hands on the sick. His skills impressed scientists at England's Royal Society. The ancient belief that touch or divine power can cure illness is supported by modern-day healers, who claim to transmit spiritual energy. To be cured, the sufferer must believe in the process.

Mary Baker Eddy (1821-1910), founder of the Church of Christ, Scientist, believed that her injuries from a fall were cured by God. Christian Scientists reject orthodox medicine – sickness is said to be a symptom of being out of touch with God. Clairvoyant **Edgar Cayce** (1877-1945) became famous for his ability to diagnose illness without meeting the patient. Miraculous cures at sites of religious visions, such as **Lourdes** in France, are investigated and given official status by the Roman Catholic Church.

Folk remedies 5

Traditional cures used readily available materials. The efficacy of some folk remedies has since been proved.

◆ **Brimstone (sulphur) and treacle** for constipation.

◆ **Brown paper soaked in vinegar** for a headache (as in *Jack and Jill*).

◆ **Cobwebs** help to heal cuts – scientists later discovered that they contain a naturally occurring antibiotic.

◆ **Dock leaves** relieve nettle stings.

◆ **The king's touch** for scrofula (tuberculosis in the neck glands).

◆ **Onions** for fever – they contain sulphur compounds, which improve the circulation and prevent disease.

◆ **Porridge** applied to the skin for acne – porridge is a natural exfoliant, and oats contain vitamin E, which is known to promote skin healing.

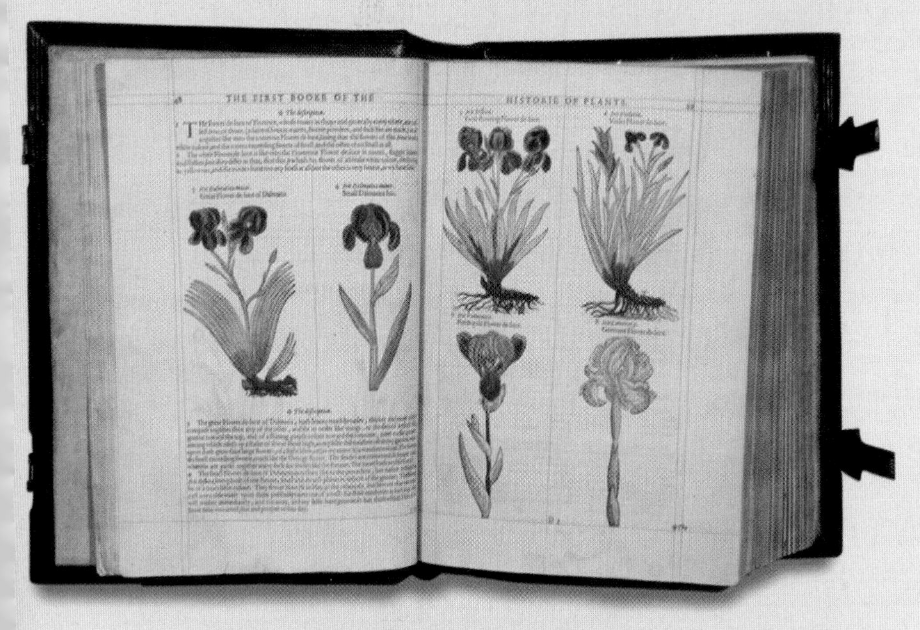

DETAILED RECORDS In *The Herball, or general histoire of plantes*, John Gerard describes several varieties of iris, a plant once used as an emetic and purgative.

Herbalism

❻

Around one in seven medical prescriptions are for plant-based drugs, echoing a long tradition of herbal remedies. The Greek physician **Dioscorides** described more than 500 in *De Materia Medica* (*c*.AD77), and the Romans spread herbalism by planting medicinal gardens in conquered lands. The tradition took root in England, where **John Gerard** (1545-1612) and **Nicholas Culpeper** (1616-54) wrote herbals (books of remedies) which became standard texts.

In the Middle Ages, monks planted medicinal herb gardens, and many monasteries ran hospitals. Apothecaries catered for townspeople who needed more than a folk remedy but couldn't afford a physician. The Worshipful Society of Apothecaries (founded in London in 1617) established the **Chelsea Physic Garden** for medicinal plants in 1673.

Many ancient remedies are still in use. Herbalists treat both physical and psychological ailments. For example, anxiety is treated with **lemon balm** or **lime tree flowers**; burns with **lavender** or **marigold**; colds with **rosemary** or **thyme**; poor digestion with **basil** or **fennel**; and immunity is boosted with **echinacea** or **garlic**.

Wise women

❼

In ancient cultures, women were healers – female doctors were common in Greek and Roman society. But after the Dark Ages, patriarchal ruling powers became suspicious of the knowledge of plant cures possessed by 'wise women'. In 1322, a Parisian court ruled that French wise woman **Jacoba Felicie** should no longer receive payment for her services. In England women were banned from practising medicine. The *Malleus Maleficarum*, a German theological treatise of 1486, condemned midwives as a danger to the Catholic Church and labelled wise women 'witches'. It was not until 1849, when **Elizabeth Blackwell** graduated as America's first doctor, that the field of medicine was reopened to women.

795

Herbal hero

English physician **Nicholas Culpeper** (1616-54) provoked the wrath of his colleagues in 1649 when he made the official College of Physician's pharmacopoeia accessible to all by translating it from Latin into English. His most influential book, *The English Physitian* (1653), later reissued as *The Complete Herbal* and still in print, set down the foundations of herbalism across the English-speaking world.

PLANT PIONEER Culpeper's herbal described 'such things only as grow in England, they being most fit for English bodies'.

RESTFUL REMEDY A 14th-century manuscript from Italy depicts the gathering of dill, traditionally used to aid digestion and encourage sleep.

Traditional medicine: East

QUESTION NUMBER

The numbers or star following the answers refer to information boxes on the right.

ANSWERS

QUESTION NUMBER	ANSWERS
5	**Acupuncture** ❷ ❺
35	**Yoga** (they are different asanas) ❼
55	**Moxibustion** ❷
76	**The feet** ❸
174	**Ayurveda, or Ayurvedic medicine** ❶ ❹
★ 227	**Zen**
228	**Meridians** ❷
354	**Ginseng** ❻
478	**Transcendental meditation** ❼
522	**D: Union** ❼
530	**B: A film with Jane Fonda** – released 1978
611	**T'ai Chi** ❽
620	**Acupressure** ❸
641	**B: *c.*1500 BC** ❶ ❹ ❺
642	**A: *c.*2500 BC** ❷ ❺
690	**True** (ephedrine, expands bronchial air passage) ❻
758	**Yin and yang** ❶ ❽
808	**Acupressure** ❸

Core facts ❶

◆ **Chinese** and **Indian** medicine date back far longer than Western, and both are widely practised today. Both use herbal remedies and physical and mental exercise to promote health.

◆ **Chinese medicine** aims to balance two opposing but complementary forces: the dark, female, moist, passive **yin**, and the bright, male, dry, assertive **yang**. An imbalance obstructs the flow of *qi* or ***chi*** (life-force).

◆ **Indian Ayurvedic medicine** is based on Hindu philosophy, and diagnoses ailments by examining the balance of three life-forces known as the **doshas**. Doctors prescribe mineral and herbal medications and perform surgery.

Acupuncture ❷

Chinese medicine teaches that the **life-force** (*qi* or *chi*) flows through the body along **14 meridians**, or channels, associated with the organs. Inserting fine needles into the skin at points along these meridians is said to restore balance by clearing blockages that impede the flow of *chi* in the body. Traditional 7th-century Chinese acupuncture charts show 365 points – modern charts show up to 2000.

Moxibustion is a variation in which the points are heated by placing cones of burning moxa (powdered mugwort leaves) on the skin, or on the end of acupuncture needles.

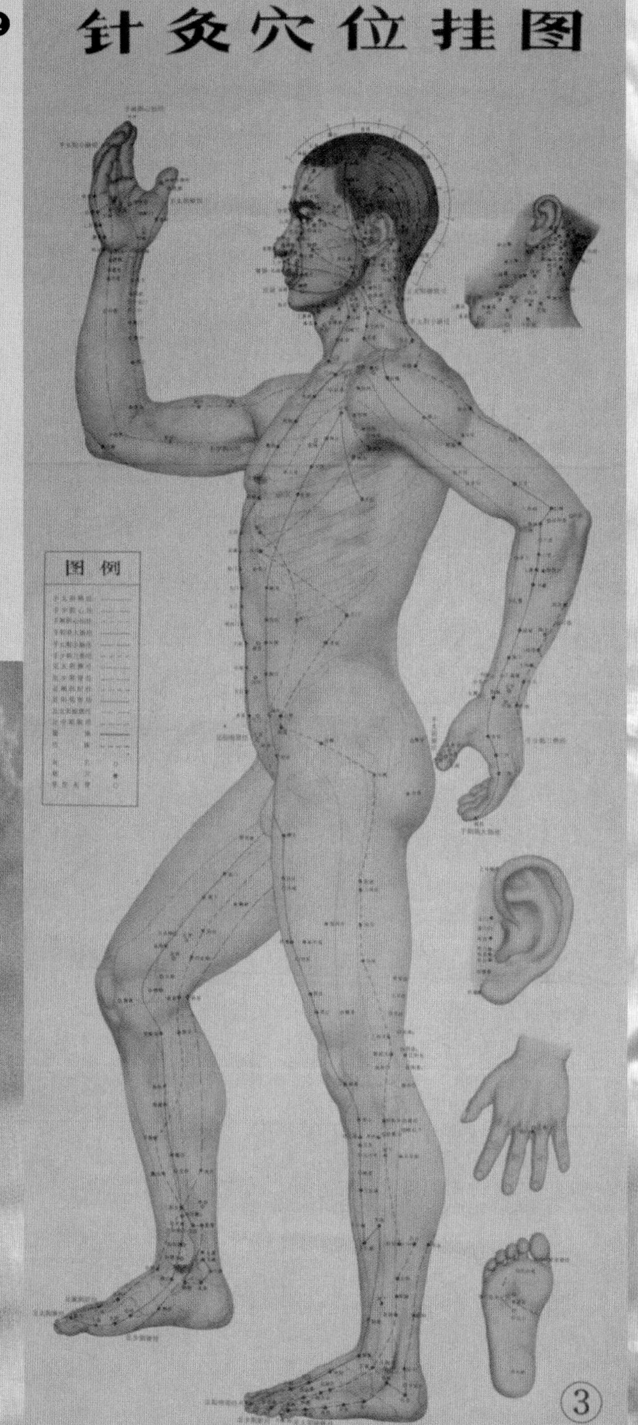

BODY MAP A 20th-century Chinese acupuncture chart shows the meridians and the points used in treatment.

Acupressure ❸

The meridian points used in acupuncture can also be manipulated by acupressure. Firm fingertip or thumb pressure is applied and massaged in the direction in which the life-force is thought to flow. To Westerners, the points used are not obviously connected with the ailment or symptom being treated – for example, travel sickness is treated by pressure on a point on the forearm about 5 cm (2 in) above the wrist.

Related techniques include Japanese **shiatsu** ('finger pressure'), which also treats points along the meridians; and **reflexology**, thought to have originated in ancient China and Egypt, in which pressure points on the feet relate to organs around the body.

Ayurveda

4

Ayurvedic ('life-knowledge') medicine is based on two main texts – the *Caraka-samhita* by a physician called Caraka, and the *Susruta-samhita* by the surgeon Susruta, which describe diseases and detailed surgical techniques.

Physicians aim to prevent disease by balancing three life-forces or **doshas**, present in everyone in varying degrees: ***vata*** (energy/the nervous system), ***pitta*** (metabolism/digestion) and ***kapha*** (body-firmness/fluid balance). Herbal remedies, dietary changes and recognising repressed emotions are believed to bring the body into balance.

DIVINE CUP Dhanvantari, Hindu physician of the gods and creator of Ayurveda, holds the elixir of immortality.

5

TIMESCALE

▶ **c.2800 BC** Mythical Chinese emperor Shen Nung is said to have invented medicine; writes herbal describing 365 medicinal plants.

▶ **c.2500 BC** Stone acupuncture needles used in inner Mongolia – later discovered by archaeologists.

▶ **c.1500 BC** Ayurvedic philosophy developed in India.

▶ **9th century BC** Ayurvedic *Caraka-samhita* and *Susruta-samhita* compiled.

▶ **3rd century BC** *Nei ching* tract on Chinese medicine completed.

▶ **1578** Scholar Li Shizen completes the comprehensive Chinese pharmacopoeia *Pen-ts'ao Kang-mu*.

Eastern herbal medicine

6

◆ Ayurvedic herbs are divided into six tastes (examples are given in brackets): **astringent** (rosemary), **bitter** (turmeric), **pungent** (black pepper), **salty** (kelp), **sour** (hawthorn berries) and **sweet** (fennel). Their properties, such as warmness or coolness, balance the doshas.

◆ Chinese herbs target body functions. For example, **dong quai** *(Angelica sinensis)* is a reproductive tonic for women, **ma huang** (*Ephedra sinica*) is a decongestant and **ginseng** *(Panax quinquefolius)* boosts energy levels to remedy nervous exhaustion.

HEALING FRUIT Chinese practitioners use aniseed-flavoured star anise (background) to treat digestive problems and improve libido.

Yoga

7

In Sanskrit, yoga means **union**, and is a process of uniting the soul with the infinite. It is a branch of the Hindu religion founded by philosopher **Patañjali**, who gathered together traditional disciplines in the *Yoga-sutras* (c.2nd century BC).

Yoga involves physical, mental and moral control and devotional duties. The physical aspect, **Hatha yoga**, is practised separately as a way of developing health, flexibility and mental discipline. It consists of **40 asanas**, or postures, that tone and stretch the body, combined with **pranayama**, or rhythmic breathing.

MENTAL FOCUS Yoga asanas evolved to aid concentration during meditation. Ancient yogis knew only one: the lotus position.

T'ai Chi

8

Chinese monks in the 5th century practised exercises based on animal movements. By the 11th century, this had been combined with Taoist principles of self-mastery and martial arts to create T'ai Chi ch'uan or 'supreme ultimate fist' ('fist' meaning exercise). Movements encourage inward focus on mental states. The aim is to balance the forces of gentle, yielding **yin** (female) and active, creative **yang** (male).

MEDITATIVE MOVES Shanghai citizens join together in T'ai Chi.

227

Zen meditation

Japanese Buddhists practice different forms of Zen meditation. **Soto** involves zazen – sitting meditation – in the lotus posture. The aim is to clear extraneous thoughts by counting the breaths. **Rinzai** involves meditating on the meaning of insoluble riddles such as 'What is the sound of one hand clapping?' The goal is enlightenment.

QUESTION NUMBER

The numbers or star following the answers refer to information boxes on the right.

ANSWERS

Question number	Answer
6	**Aromatherapy** 2
17	**Chiropractic** 4
39	**Hypnosis** 4
★ 56	**Homeopathy**
150	***Lorenzo's Oil*** – he has adrenoleukodystrophy (ALD)
389	**False** (the opposite is true) ★
479	**Bach flower remedies** 4
587	**Crystal therapy** 2
588	**Sigmund Freud** 4
628	**Actor** 2
656	**Stress** 4
665	**Osteopathy** 4
707	**Naturopathy** 4
818	**Rudolf Steiner** 2
819	**Eyesight** 4
828	**Autosuggestion** 4
829	**Anthroposophy** 2
843	**Physiotherapy** 3
897	**Aromatherapy** 2
905	**Cranial osteopathy** 4
974	**Biofeedback** 4
975	**Visualisation** 4

Other therapies

Core facts

1

◆ An **alternative therapy** is a system of treating health disorders which lies outside the boundaries of orthodox medicine.

◆ A **complementary therapy** can work alongside conventional medical treatment, and includes orthodox specialist skills such as **physiotherapy**.

◆ The effectiveness of alternative therapies is not always accepted by conventional medical practitioners. **Chiropractors**, **homeopaths** and **osteopaths** are registered and regulated by official societies; **hypnotherapists** remain unregulated, but reliable therapists can provide proof of extensive training.

Alternative and complementary therapies

2

Alternative therapies have been advocated either as complete systems of medical treatment or to deal with specific collections of conditions. The majority have not been tested scientifically, and their acceptance by conventional medical practitioners varies widely.

Alexander technique Australian actor F.M. Alexander (1869-1955) originated a technique for improving posture when he began to lose his voice on stage. Studying himself in a mirror, he found that his body slouched every time he began to speak his lines. By training himself to stand and move more naturally, he improved his breathing and speaking and also his general health. The method is claimed to boost physical and mental health and stress resistance.

Anthroposophical medicine A system of holistic (whole-body) therapy developed in the 1920s, created by followers of the philosophical ideas of Rudolf Steiner (1861-1925), better known as an educationalist. Therapists combine herbal, homeopathic and conventional remedies with eurhythmy – a series of rhythmic movements performed in time with spoken words.

Aromatherapy The use of scented plant essences, or essential oils, in massage dates back at least 1000 years, and possibly to ancient Egypt, Persia, India and China. Concentrated oils are very strong, and cannot be applied directly to the skin.To create a massage oil, a few drops of essence are added to a base oil such as sweet almond. The oils are also inhaled or used in baths or compresses; some can be taken orally, but only under a therapist's supervision.

Art therapy Swiss psychiatrist Carl Jung (1875-1961) first used painting and modelling to enable self-expression in his patients. During World War II, an ex-student of Jung's, Irene Champernowne, with artist Adrian Hill, developed the technique into a stand-alone therapy. Its value is now widely accepted by medical practitioners.

MYSTICAL CURES Practitioners of crystal therapy believe that the electromagnetic energy of crystals gives them healing properties.

Physiotherapy

3

Physiotherapists treat physical injuries, rehabilitate patients after an illness or operation, and maintain mobility and strength in those suffering long-term illness. Basic techniques include **exercise** and **massage** – for example, lung congestion in cystic fibrosis is treated by tapping the chest with the fingers. The application of heat and/or cold using **hydrotherapy** (water) or **ultrasound** (which heats targeted areas of tissue) is used to relieve pain and swelling and aid blood circulation, and severe mobility problems can be treated with **electrical muscle stimulation**.

QUICK RELIEF Physiotherapists can specialise in the treatment of sports injuries such as muscle strains and torn ligaments, and are often required to provide instant therapy on the field.

Therapies (continued)

Autogenic training In the 1920s, German neurologist Johannes Schultz (1884-1970) originated a system of deep relaxation based on his study of hypnosis. Training involves the repetition of phrases designed to induce a slowing of the heartbeat, regular breathing and the dilation of blood vessels, leading to feelings of warmth and heaviness. Investigations suggest that it is an effective stress-reliever, and it has been used to treat insomnia and migraine.

Autosuggestion Created by French pharmacist Emile Coué (1857-1926), autosuggestion (also known as Couéism) is a form of meditation with elements of self-hypnosis. Patients repeat phrases such as 'Every day, in every way, I am getting better and better' to empty the mind and create positivity. His ideas have been incorporated into therapies such as autogenic training and visualisation.

Bach remedies British homeopathic and orthodox physician Dr Edward Bach (1880-1936) developed a series of 38 flower-based oral remedies and a combination 'rescue remedy' for his patients by floating flowers on spring water in sunlight to 'infuse' the water with their essence. The remedies treat the whole person, rather than a specific ailment.

Bates method This series of eye exercises developed by New York ophthalmologist William Bates (1860-1931) aims to improve eyesight and avoid or reduce the need for glasses. It does not claim to help conditions such as glaucoma or cataracts, which need conventional professional attention.

Biofeedback Experiments have shown that people can exercise a degree of control over normally unconscious bodily functions governed by the autonomic nervous system, such as blood pressure and skin temperature. In biofeedback, electronic sensors and read-out devices show what is happening in the body, so that the subject can learn to recognise and alter these responses and thus influence anxiety and raised blood pressure.

Chiropractic Chiropractic is a manipulation therapy similar to osteopathy, but closer to conventional medicine because it uses X-rays and other conventional methods of diagnosis. ❹ Its originator, Canadian healer Daniel David Palmer (1845-1913), first used the technique in 1895. Chiropractic successfully treats disorders caused by musculo-skeletal misalignment – from back pain to hearing problems – using hand pressure, thrusts and rotation to take a 'locked' joint slightly beyond its present range and free it.

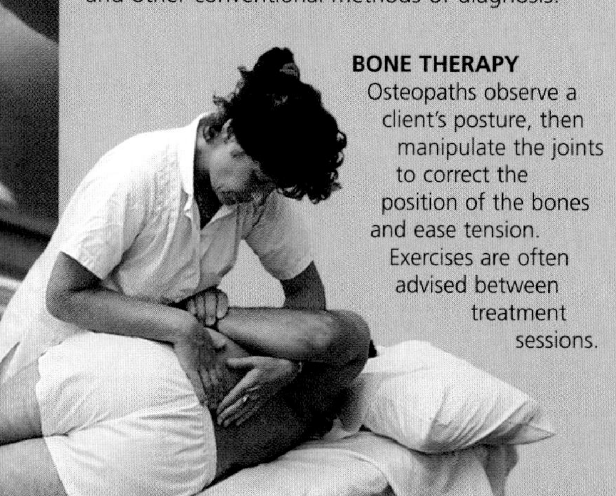

BONE THERAPY Osteopaths observe a client's posture, then manipulate the joints to correct the position of the bones and ease tension. Exercises are often advised between treatment sessions.

Feldenkrais method Russian-Israeli engineer and sportsman Moshe Feldenkrais (1908-84) taught efficient patterns of movement through manipulation – for example, by rotating the upper and/or lower torso and the shoulders and pelvis. These manipulations claim to develop corresponding new patterns in the brain which help to break bad movement habits that cause the body discomfort.

Flotation therapy Floating in water in the dark in an enclosed tank is claimed to relieve stress-related disorders. The tank is filled up to 25 cm (10 in) deep with skin-temperature water made buoyant with a high concentration of dissolved salts. Skilled supervision is vital – the sensory deprivation that occurs may worsen psychiatric conditions such as depression.

Hypnotherapy Hypnotherapists induce a trance state in clients. In this state, the client can be regressed into past events to provide an insight into current behaviour, and given suggestions about how to improve reactions to problems. Modern practice began with Franz Mesmer (1734-1815), who called hypnotism 'animal magnetism'. French neurologist Jean Charcot (1825-93), with whom Freud studied, pioneered its use in treating mental conditions. Today it is used for certain stress-related conditions and to help people stop smoking.

Naturopathy According to naturopaths, identifying the underlying cause of illness is the key to good health. Naturopathy, originally known as 'nature cure', works on the principle of helping the body to heal itself. A combination of diet, exercise, massage, hydrotherapy, relaxation, chiropractic and osteopathy are used to promote health. Naturopaths aim to strengthen the body's own disease-resistance and curative mechanisms.

Osteopathy In the 1870s, American doctor, Andrew Taylor Still (1828-1917) founded the practice of osteopathy, now one of the most widely used forms of alternative and complementary therapy. It is a manipulation and massage therapy used for treating pain and injuries affecting the bones, joints, muscles and ligaments, and is also used to treat tension headaches and some forms of arthritis. Many doctors recognise its benefits, but most dispute its value for non-mechanical disorders.

Visualisation therapy The systematic creation of positive images in the imagination is claimed to help overcome negative behaviour and improve bodily health. Claims that it can treat diseases such as cancer – for example, by visualising the body's cells attacking the tumour – are controversial, but a positive mental attitude is known to help sufferers of serious illness cope with their symptoms.

56 ★

Homeopathy

German physician **Samuel Hahnemann** (1755-1843) developed **homeopathic medicine** in about 1810. Hahnemann noticed that chinchona bark, a herbal remedy for malaria which contains quinine, produces headaches and fever that mimic malaria symptoms. From this he developed the doctrine of 'like curing like' – substances that cause symptoms similar to those of a particular disease can help to cure the disease. Remedies are highly diluted and are designed to stimulate the body to heal itself.

ALTERED STATES Colour therapy uses coloured light on the body to influence moods and emotions.

WEIRD AND WONDERFUL ❺

Kirlian photography claims to capture a person's **aura** on film as a coloured halo around the head or hands. Practitioners believe that health disorders can be diagnosed before symptoms arise by examining the aura.

QUESTION NUMBER

The numbers or star following the answers refer to information boxes on the right.

ANSWERS

33 **A cow** (*vacca* in Latin) ❶❸❺❻

156 **Meningitis** (as yet no vaccines for the others) ❺

175 **Smallpox** ❸

176 **Sewers** ❷

203 **Meningitis** ❺

220 **Breast cancer and cervical cancer** ❹

246 **The BCG vaccination against TB** ❻

372 **Louis Pasteur** ❸

428 **Poliomyelitis** ❺❼

432 **Rabies** (in 1885; polio vaccine 1954) ❸❻

433 **1977** (accept 1975-9) ❻

470 **Poliomyelitis** ❻

792 **Sewers** ❷

★ **798** **Cholera**

809 **Louis Pasteur** ❸

898 **Vaccination** ❸

976 **The Great Stink** ❷

991 **Typhoid** ❺❻

Disease prevention

Core facts ❶

◆ The greatest advances in reducing death rates and improving health have come from **preventing disease** rather than curing it.
◆ Public health and **hygiene** campaigns during the 19th century – the first effective steps in disease prevention – led to better sanitation and housing and clean water supplies.
◆ The most effective measure against individual diseases is **vaccination**. An understanding of the immune system has made it possible to create vaccines for newly emerging diseases.
◆ **Health campaigns** raise public awareness about disease prevention, using literature, television advertising and electronic mail.

Health campaigns ❷

In the 19th century, residents in poor urban areas suffered squalid living conditions. Overcrowding, lack of ventilation and light and open drains contributed to epidemics of **cholera** and other diseases.

In England, campaigning physician **Edwin Chadwick** (1800-90) instigated the Public Health Act of 1848, which made hygiene a local responsiblity – though it did not stop 11 000 deaths in a cholera epidemic in 1853-4, or the **Great Stink** of 1858 when summer heat made the Thames smell like an open sewer. But within 30 years, London's cesspits and easily contaminated wells had been replaced by a **sewer network** of glazed earthenware pipes and **mains water supplies** – an example eventually followed by major cities worldwide.

LONDON'S CLEAN-UP A pipe (centre) carries sewage beneath Charing Cross Station. Engineer Joseph Bazalgette supervised the construction of 1600 km (1000 miles) of sewers leading to a main, low-level pipe running along the Thames Embankment.

★ 798

Cholera

In London's cholera epidemic of 1853-4, 700 people died in one two-week period in Soho. They all lived near a **water pump** close to a leaky cesspool. Local resident **Dr John Snow** believed that water from the pump transmitted the disease. He persuaded officials to remove the pump handle, and the epidemic ended within days.

A MEAGRE LIVING A street stall in Houndsditch, drawn by French artist Gustave Doré, illustrates the awful conditions for London's poor in the 1870s.

Vaccination

Chinese physicians inoculated children against smallpox in the 11th century by placing the scabs of people with a mild form of the disease into the nostrils. This technique – **variolation** – appeared in Britain and America in 1721, but risked causing full-blown smallpox.

In 1796, British physician **Edward Jenner** (1749-1823) performed the first vaccination, infecting a child with cowpox (a mild disorder), then with **smallpox** – the cowpox infection prevented smallpox from taking hold. Using the same principles, French chemist **Louis Pasteur** (1822-95) created vaccines against **anthrax** (1881) and **rabies** (1885).

The ability of vaccines to provoke antibody production in the immune system was not understood until the mid-20th century. Vaccines are now made using **attenuated** (weakened) disease organisms or **antigens** (protein molecules) from disease organisms, or are **genetically engineered**.

LIFE-SAVER Pasteur used weakened strains of bacteria to develop protective injections.

NATURAL DEFENCE Jenner noticed that dairymaids who caught cowpox (below) were immune to smallpox.

Health screening

Preventive medicine includes screening to detect disease before symptoms appear, or to evaluate the risk of developing disease. Tests are used only if effective treatment can be given.

Routine tests include: blood testing newborns for **metabolic disorders** such as hypothyroidism; cervical and breast **cancer** checks for women; measurements of **blood pressure** and **cholesterol levels** to predict the likelihood of heart disease; and **eye pressure** tests to detect glaucoma.

Major vaccines

◆ **Bacillus Calmette-Guérin (BCG)** Single tuberculosis vaccination given to those at risk at age 12-13; gives up to 80 per cent protection for 10 years.
◆ **Cholera** Two injections a week apart, giving varying degrees of immunity for 6 months; new versions in development.
◆ **Haemophilus influenzae B (Hib)** Three injections between 2 and 6 months of age; booster at 18 months. Protects against bacterial meningitis.
◆ **Triple vaccine** Diphtheria, tetanus and pertussis (whooping cough) vaccine given in three doses from 3-14 months of age with booster at five years old.
◆ **MMR** Measles, mumps and rubella injection given at two years old.
◆ **Poliomyelitis** Taken orally in liquid form during childhood.
◆ **Typhoid** Two doses, four to six weeks apart, giving protection lasting 1-3 years.

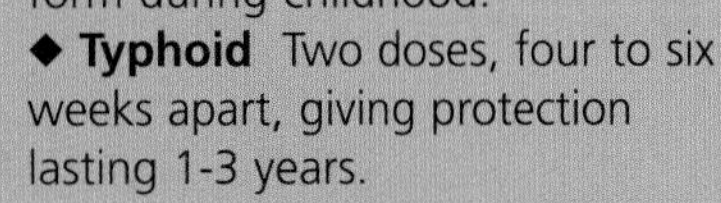

TIMESCALE

▶ **430 BC** Greek historian Thucydides notes that plague does not infect people twice.
▶ **11th century** The Chinese use variolation to prevent smallpox.
▶ **1796** Edward Jenner's first smallpox vaccination.
▶ **1870** Louis Pasteur produces first live vaccine.
▶ **1882-3** German doctor Robert Koch discovers the bacteria that cause tuberculosis and cholera.
▶ **1896** English scientist Almroth Wright produces typhoid vaccine.
▶ **1921** French scientists Albert Calmette and Camille Guérin first use BCG vaccine.
▶ **1954** American Jonas Salk's polio vaccine.
▶ **1957** Russian-born Albert Sabin improves on Salk's polio vaccine.
▶ **1971** MMR first used.
▶ **1977** Last recorded smallpox case worldwide.
▶ **1986** First genetically engineered vaccine.
▶ **1986** First acellular vaccines, using only the antigenic parts (proteins) of disease organisms.

Modern awareness

Governments in developed countries work with local and national health authorities and charities to publicise disease prevention. Campaigns provide information on issues such as HIV/AIDS prevention, antibiotic overuse, flu immunisation, the risks of obesity, the effects of smoking, and on spotting symptoms of conditions such as cancer. The UN's **World Health Organisation (WHO)**, founded in 1948, works towards establishing preventative medicine worldwide. One of its current campaigns is to eradicate polio – this has brought down the number of cases worldwide to fewer than 700 in 2003.

CLEAN WATER Basic sanitation and clean water supplies reduce disease in developing countries.

Surgery 1

QUESTION NUMBER

The numbers or star following the answers refer to information boxes on the right.

ANSWERS

QUESTION NUMBER	ANSWERS
4	**Anaesthetic** – sometimes used in amputations
10	**Amputation** – to prevent infection spreading
14	**Tourniquet** ('a turning instrument', or swivel) 4
106	**True** 6
177	**Trepanning** 2 3 5
247	***Chirurgia*** (modern French *chirurgie*) 2
348	**Chloroform** 5
380	**Joseph Lister** 6
437	**About 1500** 4
440	**Cataract** 5
491	**True** 2
591	**Boiling oil** 4
★ 600	**Blood and bandages**
643	**B: 1736** 5
644	**D: 1884** 5
830	**Aseptic surgery** 1 6
844	**Skin graft** 2
959	**D: Nitrous oxide** (laughing gas) 5
960	**B: Blood groups** 6

Core facts 1

◆ **Surgery** is the use of instruments to operate on parts of the body – removing, repairing, altering or replacing diseased, injured or malfunctioning organs and parts.
◆ Surgeons formed a separate branch of the medical profession from the late Middle Ages. Until the late 19th century, surgery carried risks of shock, infection and heavy blood loss – now controlled by **anaesthesia**, **asepsis** (sterile operating conditions) and **blood transfusions**.
◆ In the 20th century, technological advances led to the development of **keyhole** techniques, **laser surgery** and **microsurgery**.

Ancient and medieval 2

Evidence of **amputation** – cut marks on an arm bone found in Iraq – dates back 45 000 years. In the 6th century BC, Hindu surgeon **Susruta** used **skin grafts** to reconstruct damaged facial tissue. His knowledge passed to the Arabs, Greeks and Romans.

Early European surgeons consulted the first treatise on surgical techniques, written by **Abu al-Qasim** (*c*.936-*c*.1013) from Muslim Spain. French surgeon **Guy de Chauliac's** *Chirurgia Magna* (Great Surgery) of 1363, describing procedures such as hernia and cataract operations, later became a standard text.

SECOND-CLASS SKILL Surgery – including trepanning (right) – was seen as a lowly occupation compared to the medieval physician with skills of diagnosis and treatment, seen below in a 15th-century copy of the *Chirurgia Magna*.

WEIRD AND WONDERFUL 3

Ancient skulls with holes bored into them are examples of a process known as **trepanning** – probably intended to relieve headaches or release evil spirits. Signs of healing on the skulls proves that patients survived.

 600

Barber-surgeons

In the Middle Ages, barbers performed minor surgery, advertising their skills with buckets of blood and bandages outside their premises (the origin of the red and white **barber's pole**).

In Britain, a hierarchy of apprenticed surgeons, barber-surgeons and barbers emerged, and founded the **United Barber-Surgeon Company** in 1540. In 1745, the surgeons formed their own company, which became the **Royal College of Surgeons** in 1800.

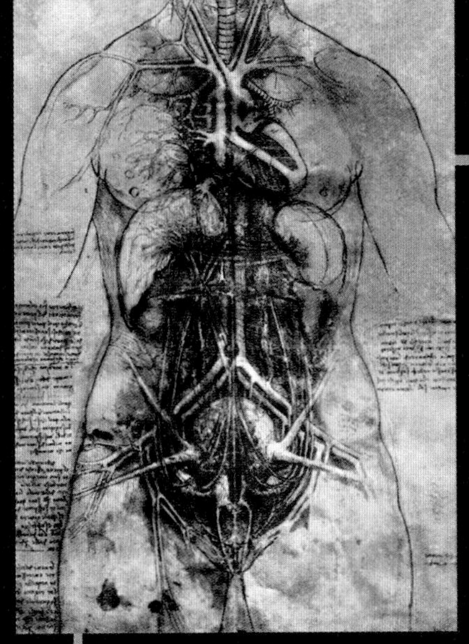

BODY BLUEPRINT Da Vinci died before gathering his notes into a planned anatomical treatise.

Renaissance and beyond

Leonardo da Vinci (1452-1519) carried out 30 human dissections, observed many more in Milan, Florence and Pavia and used his experiences to produce accurate drawings of the human body. Belgian physician **Andreas Vesalius** (1514-64) worked from dissections to produce the first anatomy textbook, *De Humani Corporis Fabrica* (On the Structure of the Human Body), in 1543.

French surgeon **Ambroise Paré** (c.1510-90), the 'father of modern surgery', developed new principles based on practical experience. The traditional use of boiling oil to cauterise gunshot wounds caused inflammation and pain – Paré ran out of oil at the Siege of Turin, and used a mixture of egg yolk, rose oil and turpentine instead, with better results. He also introduced the technique of clamping and tying arteries to stop them bleeding. **Jean-Louis Petit** (1674-1750), the first director of the French Academy of Surgery, improved military surgery by inventing a screw tourniquet that made amputation safer.

In Britain, **William Cheselden** (1688-1752) became known for his lithotomies (removing kidney or bladder stones), performed in less than a minute. He taught Scottish surgeon **John Hunter** (1728-93), whose rigorous approach to surgery, based on biological principles, led to its recognition as a science.

HIGH-SPEED AMPUTATION Lack of proper anaesthetics forced a good 18th-century surgeon to be quick as well as accurate.

TIMESCALE

- ▶ **c.8000 BC** Trepanning widely practised.
- ▶ **10th-12th centuries** Surgeons in Muslim Spain operate on cataracts and remove kidney stones.
- ▶ **13th-14th centuries** Key Arabic surgical texts translated into Latin.
- ▶ **1363** *Chirurgia Magna* by Guy de Chauliac.
- ▶ **1536** Paré develops a new way of cauterising gunshot wounds at the Siege of Turin.
- ▶ **1543** Vesalius's *De Humani Corporis Fabrica*.
- ▶ **1736** English surgeon Claudius Amyand performs the first appendicectomy.
- ▶ **1844-7** Nitrous oxide used as anaesthetic, then ether and chloroform.
- ▶ **1865** Joseph Lister introduces concept of antiseptic surgery.
- ▶ **1884** Cocaine used as local anaesthetic.
- ▶ **c.1915** British surgeon Harold Gillies pioneers plastic surgery.
- ▶ **1952** Microsurgery on middle ear.
- ▶ **1963** Laser surgery invented.
- ▶ **1967** First human heart transplant (in South Africa).
- ▶ **1980s** 'Keyhole' surgery developed.

A new century

Medical innovations in the 19th century made surgery a routine method of treatment. **Anaesthetics**, introduced in the 1840s, prevented pain, allowed more time for surgical procedures and reduced the risk of shock. In 1865, English surgeon **Joseph Lister** (1827-1912) introduced surgical hygiene, using carbolic acid (phenol) as an **antiseptic** to sterilise surgical instruments and wounds. He reduced death rates after surgery from up to 50 per cent to 2-5 per cent.

Two further advances came at the end of the century: **X-rays**, first used in 1895, enabled imaging of the inside of the body, and the discovery of **blood groups** in 1900 made vital blood transfusions safe.

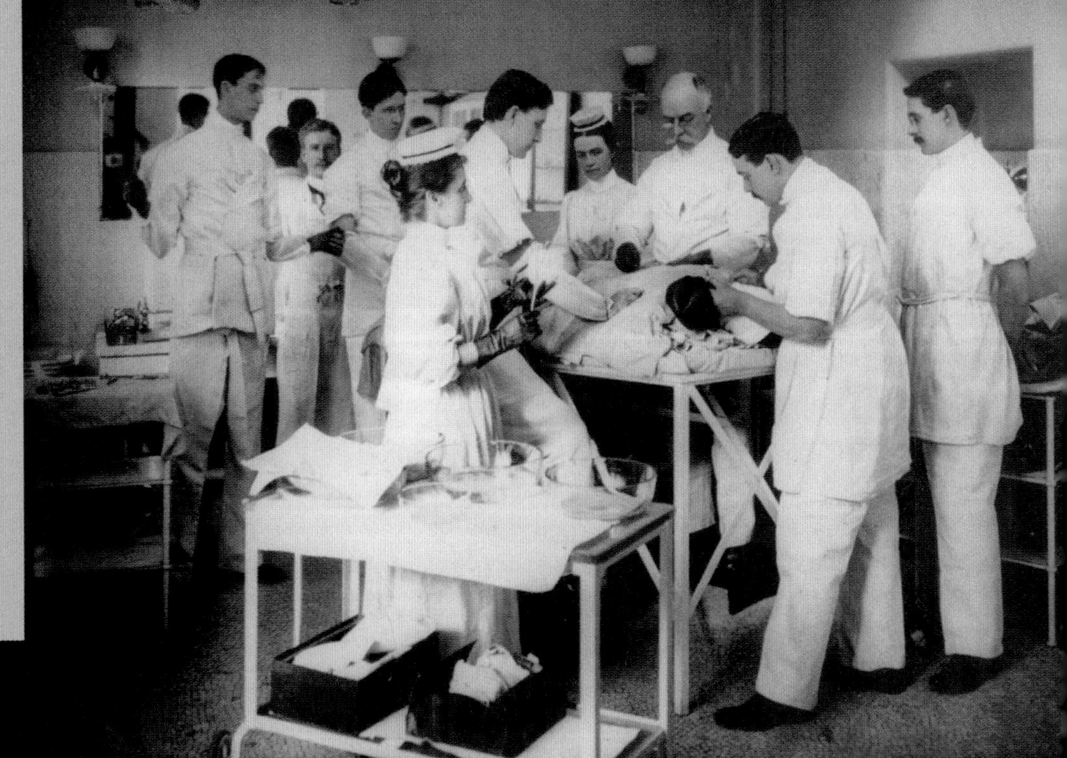

ENTERING THE MODERN AGE A New York operating theatre of 1900 used local anaesthetics and sterile equipment.

Surgery 2

QUESTION NUMBER

The numbers or star following the answers refer to information boxes on the right.

ANSWERS

12 **Local anaesthetic** 4

18 **Keyhole surgery** 5

29 **True** 4

83 **Intensive care unit** – constant patient monitoring

153 **Thermometer** (others can all be used in surgery) 2

178 **Microsurgery** 5

206 **Laser** 5

287 **False** (for slowly infusing fluid into the body) 2

311 **George Clooney, or Alex Kingston** – started 1994

574 **Scalpel** 2

578 **Clamping forceps** 2

579 **Dissecting forceps** 2

629 **Anaesthetist, or anaesthesiologist** 1

666 **Anaesthetic** 4

667 **Keyhole surgery** 5

722 **Scrubbing up** 1

★ 723 **Stitches**

841 **Operating theatre** 1

842 **Sterilising instruments** 2

847 **Scalpel** 2

880 **Removal of a kidney or bladder stone** 6

906 **Epidural, or spinal** 4

Surgery today

1

Operating theatres are sterile environments run by a skilled medical team. The **surgeon** works with assistant surgeons, an **anaesthetist** and a **scrub nurse** (who has 'scrubbed up', or washed his or her hands and arms thoroughly).

The anaesthetist is a leading member of the team, responsible for keeping the patient unconscious (or, in some cases, conscious but unable to feel pain) and monitoring life support (heart and breathing). The scrub nurse passes the surgical instruments and assists in cleaning wounds and controlling bleeding.

The major areas of modern surgery are **wound treatment** (cleaning and closing wounds), **extirpation** (the removal of an organ or large area of tissue), **transplantation** and **reconstruction** (grafting tissue to rebuild parts of the body removed by surgery or trauma).

TEAMWORK A surgeon (far right), assisted by a nurse, prepares to operate on a patient's colon using an endoscope. An anaesthetist (left) monitors the patient.

★ 723

Sutures

Sutures (stitches) close wounds, prevent bleeding and infection and allow skin to regrow together. **Adhesive tape** is used for small incisions, **needle and thread** for larger cuts, using soluble 'catgut' (sheep's intestine) or insoluble silk or nylon. **Stainless steel** is used for strong joins such as **wiring** bone together, or for **stapling** internal organs or skin wounds.

PRECISION INSTRUMENTS The most common surgical tools are scalpels and clamps.

Instruments

2

- **Autoclave** Steam-heated oven for sterilising surgical instruments.
- **Catheter** Flexible tube for feeding, draining fluids or administering drugs.
- **Clamp** A device which compresses tissue together – used to close arteries, for example.
- **Drip** A tube which feeds a continuous slow flow of fluid, such as morphine, into the patient.
- **Forceps** A gripping instrument similar to tweezers or pliers.
- **Retractor** A device for holding tissues out of the way of the area being operated on.
- **Scalpel** Small surgical knife used to cut tissues.
- **Speculum** A device which holds open a body passage such as the anus or vagina.
- **Swab** Absorbent pad used for cleaning up.

Tie-breaker

3

Q What is the literal meaning of the word 'anaesthesia'?

A 'Without feeling', from the Greek, via Latin, *an* (without) and *aisthesis* (feeling or perception).

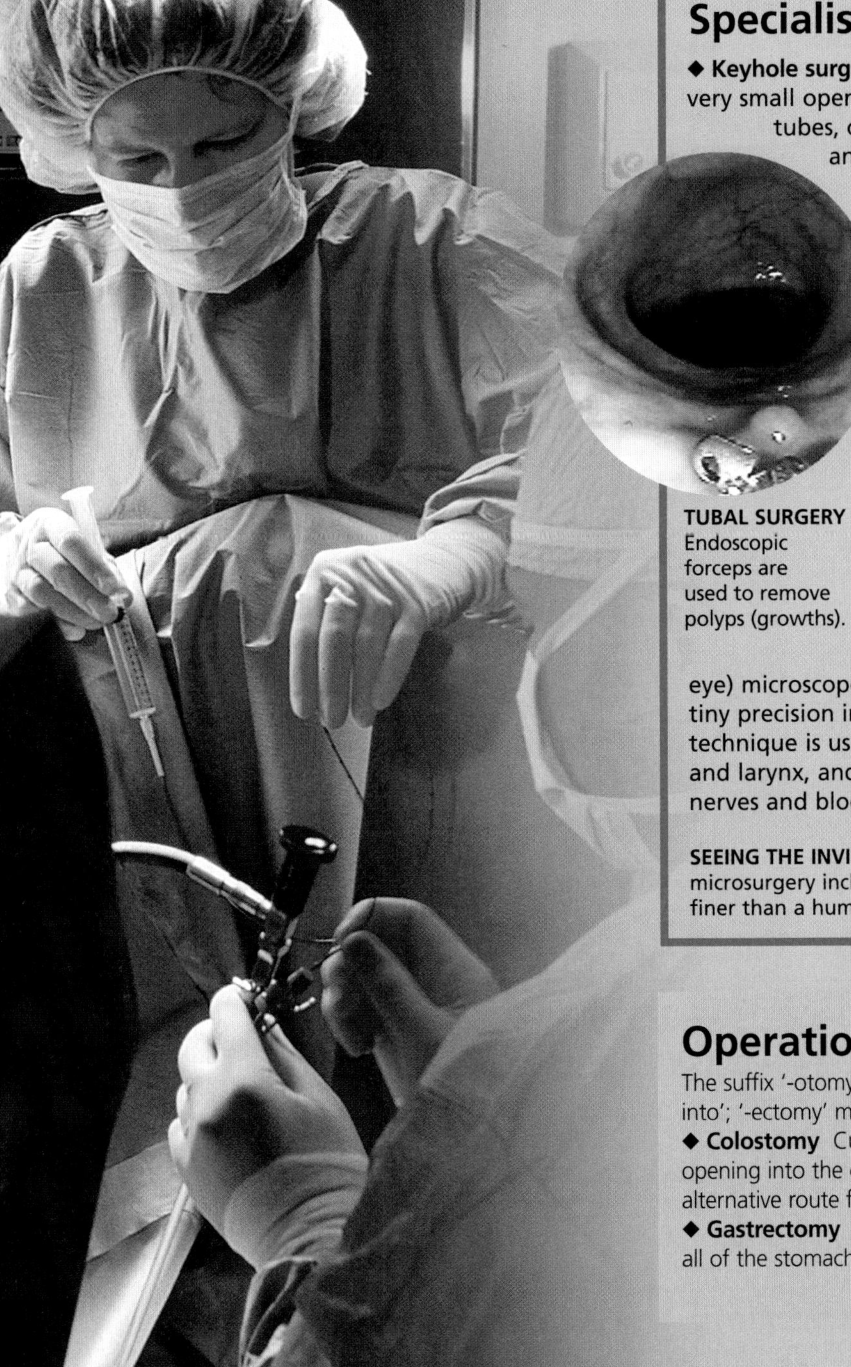

Anaesthesia

A **general anaesthetic** keeps the patient unconscious. It is initiated with an injection and maintained by administering nitrous oxide and oxygen through a breathing mask.

In many operations, the patient remains conscious but the area being operated on is numbed by injecting drugs which block the transmission of nerve impulses. A **local anaesthetic** numbs the operation site; a **regional anaesthetic** numbs a larger area of the body. Local anaesthesia can be used in brain surgery – the brain does not feel pain directly – allowing the patient to report any effects.

Acupuncture has also been used as a local anaesthetic. Needles are passed into specific points in the skin, and an electric current is passed through them. Research suggests that this stimulates the nerves to release endorphins (pain-killing chemicals).

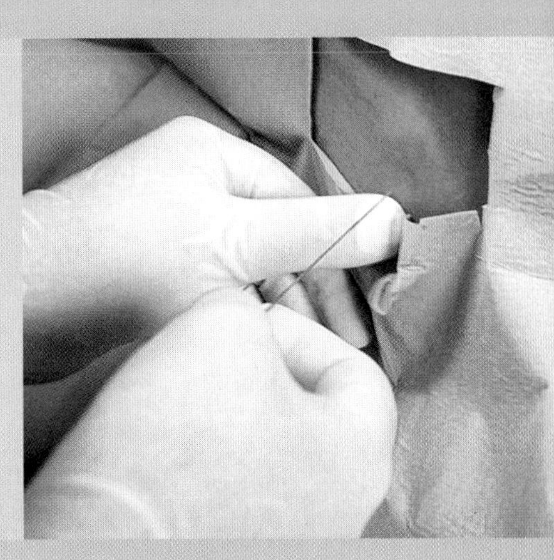

NERVE BLOCK In an epidural, anaesthetic is injected inside the membrane around the spinal cord to numb the lower body.

Specialised surgical techniques

◆ **Keyhole surgery** Minimally invasive surgery (MIS) involves making very small openings in the body, then feeding through narrow tubes, called laparoscopes, which take instruments, a light and camera to the operating site. Used for operations such as gall bladder removal or for extracting stones from the urinary tract, the technique was developed in the 1980s, when silicon chips were first used in TV cameras – the surgeon observes the operation in images transmitted by a tiny closed-circuit camera via fibre optic cables.

◆ **Laser surgery** The light from laser beams can be used to cut soft tissue, coagulate blood and seal blood vessels, or encourage tissue healing. Laser surgery is often administered through an endoscope (tube), and is used in eye surgery (for repairing detached retinas, for example), skin surgery such as birthmark removal, and to destroy abnormal cells in danger of becoming malignant.

TUBAL SURGERY Endoscopic forceps are used to remove polyps (growths).

◆ **Microsurgery** The surgeon uses a binocular (two-eye) microscope to manipulate tiny precision instruments. The technique is used on the ear, eye and larynx, and to rejoin severed nerves and blood vessels.

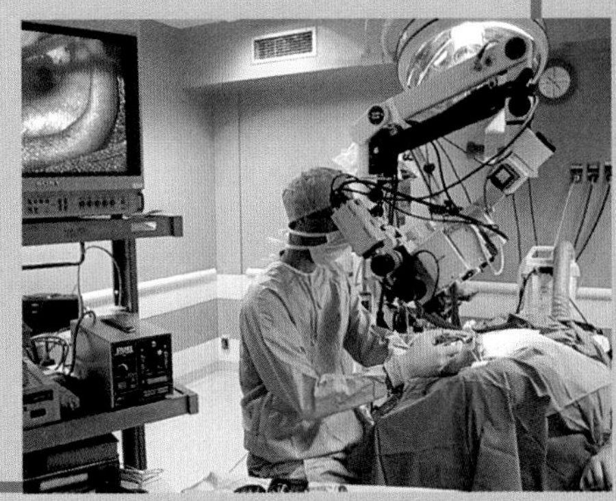

SEEING THE INVISIBLE Tools used in microsurgery include curved needles finer than a human hair.

Operations

The suffix '-otomy' means 'cutting into'; '-ectomy' means 'removal of'.

◆ **Colostomy** Cutting an artificial opening into the colon to allow an alternative route for waste removal.

◆ **Gastrectomy** Removal of part or all of the stomach.

◆ **Lithotomy** Incision into a duct or organ such as the gall bladder to remove calculi (stones).

◆ **Nephrectomy** Kidney removal.

◆ **Oophorectomy** Ovary removal.

◆ **Pneumonectomy** Removal of part or all of a lung.

◆ **Valvotomy** Widening of a narrowed heart valve.

Implants and transplants

QUESTION NUMBER

The numbers or star following the answers refer to information boxes on the right.

ANSWERS

70	**A: Artificial heart** 3 5 7
108	**True** (specimen found in France, 1997) 2
207	**Hip replacement** 3
239	**1980s** 7
320	**Zaphod Beeblebrox** – by Douglas Adams, 1979
330	**Tissue rejection** 1 6
438	**1960** (accept 1958-1962) 3
448	**Deafness** 3
492	**True** 1 4 7
★ 528	**C: An implant material**
553	**Bone-marrow transplant** 4 7
654	**Kidney** 7
668	**Hip replacement** 3
800	**Hip replacement** 5
931	**True** (cornea 1906, kidney 1954) 2 4 7
977	**Transplant is living tissue; implant is not** 1
978	**Bionic** 1 3

Core facts 1

◆ A **prosthesis** is any artificial device or machine placed in the body (or attached to it) to help or to replace the function of a body part. An **implant** is an internal prosthesis.
◆ A **transplant** (also called a graft) is living tissue that is used to replace a malfunctioning organ or body part. It may come from another part of the body (an **autograft**), or from another person (an **allograft** or homograft) or animal species (a **xenograft** or heterograft).
◆ **Bionic** devices mimic natural body functions through the use of electronics and prosthetics.
◆ A major problem of transplants is the immune system's **rejection** of 'foreign' tissues.

Replacement parts 2

Replacements for body parts lost through injury or disease have been made for centuries. Until the late 19th century they were relatively crude – mainly wooden legs, artificial hands, false teeth and glass eyes. Then the era of precision engineering and particularly microelectronics began.

Skin grafting (a skin transplant using the patient's own skin) was carried out in ancient India as part of nose reconstruction, and in Renaissance Italy. It became a routine part of plastic surgery during the two world wars of the 20th century. Meanwhile, in 1906 the **first successful allograft transplant**, of a cornea, was carried out. It was not rejected because the cornea has no direct blood supply.

Implants 3

Passive implants are mechanical devices such as steel rods and plates used to repair broken bones, plastic eye lenses and cosmetic breast implants. **Active and bionic implants** work by mechanical or electric power or are electronically operated.

PASSIVE IMPLANTS

◆ **Artificial joints** French surgeon Louis Ollier did the first elbow replacements around 1900. Now, more than 50 000 hip joints are replaced in the UK each year, plus knee and finger joints.
◆ **Blood vessels** Plastic blood-vessel sections may be used – for example, to repair a swollen or burst aorta.
◆ **Dental implants** A small post, usually of titanium, is fixed into the jaw. When this has healed, a permanent artificial tooth is attached.

ACTIVE AND BIONIC IMPLANTS

◆ **Artificial hearts** These appeared in the 1980s. They can cause blood-clotting, but may be used to keep patients alive until donor hearts are found.
◆ **Artificial limbs and hands** Simple mechanical legs date from at least the 1600s. Modern versions operate electronically, activating small electric motors in response to nerve impulses in the wearer's stump.
◆ **Bionic eye** An electronic detector is implanted in the retina of a blind person's eye to send signals to the brain along the optic nerve. One version receives images electronically from tiny video cameras worn like glasses.
◆ **Cochlear implants** Signals from a device like a hearing aid stimulate the auditory nerve.
◆ **Heart pacemakers** These regulate the heartbeat in people with heart-rhythm disorders. The first were fitted in 1960; programmable computer-controlled pacemakers were introduced in the 1980s.

ARTIFICIAL HIP The hip joints bear great stresses and prostheses tend to loosen or wear out. They are now made of stainless steel and polyethylene.

Transplants

These can save lives, but are only successful if the natural process of tissue-rejection can be overcome.

◆ **Bone marrow** In some forms of leukaemia and other bone-marrow diseases, a patient's marrow is first destroyed by chemotherapy or radiation. Then healthy marrow from a donor, or harvested from the patient when the disease was inactive, is injected into the blood.

◆ **Cornea** Transplanted corneas – the eye's transparent covering – can correct some forms of blindness.

◆ **Heart** Hundreds of donor hearts are transplanted every year – the longest-surviving recipient lived for 23 years after the operation.

◆ **Heart and lungs** Combined transplants are now a standard procedure for people with irreversible lung disease, such as cystic fibrosis with heart problems. It is technically easier than replacing either organ alone.

◆ **Heart valves** Human or pig heart valves are transplanted.

◆ **Intestines** If the small intestine is removed because of serious bowel disease, it is possible to transplant a replacement into the patient.

◆ **Kidneys** The body can function with just one healthy kidney, so kidneys are transplanted from living as well as dead donors.

◆ **Liver** This is technically difficult, and people with advanced liver failure often have other complications. But there are fewer rejection problems than for kidneys.

◆ **Pancreas** A pancreas may be transplanted, usually with the connecting segment of small intestine, into a severely diabetic patient. Alternatively, the insulin-producing islets of Langerhans from a pancreas may be injected into the patient.

VIRTUAL SURGERY
A surgeon manipulates callipers on a screen in a simulation of a liver operation (below right).

Pioneers

Christiaan Barnard (1922-2001) South African surgeon who performed the first successful heart transplant in 1967.

John Charnley (1911-82) British pioneer of hip-replacement surgery.

Macfarlane Burnet (1899-1985) and **Peter Medawar** (1915-87) Australian and British Nobel prizewinners who investigated the immune system.

Donall Thomas (1920-) American bone-marrow transplant pioneer and Nobel prizewinner.

Robert Jarvik (1946-) American physician who designed the first artificial heart.

Tissue-matching

If a donor's tissues do not match those of the recipient, they will be rejected. Tissues from a live donor (a kidney, for example) or from a brain-dead person whose heart is still beating give the best results. Even with a match, immunosuppressant drugs are needed.

MOTHER OF PEARL
A fragment implanted into diseased bone causes the bone to regenerate around it.

TIMESCALE

▶ **1906** Successful corneal transplant.
▶ **1938** Hip replacement.
▶ **1952** Successful heart-valve implants.
▶ **1954** Successful kidney transplant.
▶ **1956** Successful bone-marrow transplant.
▶ **1960s** Bionic artificial limbs developed.
▶ **1963** Liver transplant.
▶ **1966** Pancreas transplant.
▶ **1968** First successful bone-marrow transplant.
▶ **1969** Bioglass invented.
▶ **1972** First use of pig's heart valves for humans.
▶ **1981** Successful heart and lung transplant.
▶ **1982** Jarvik-7 artificial heart implanted.
▶ **1982** Multichannel cochlear implants.
▶ **c.1985-2001** Tissue-cultured and artificial skin developed.

Bioglass

Implanted materials must be compatible with body tissues. Bioglass, a type of glass made with calcium, phosphate and sodium salts, was the first manufactured material to **bond naturally with bone** and connective tissues. A liquid Bioglass containing bone cells can induce bone to regenerate faster than normal.

Medical machines

QUESTION NUMBER

The numbers or star following the answers refer to information boxes on the right.

ANSWERS

53 **Defibrillator** – corrects erratic heart rate (fibrillation)

★ **57** **Pain**

107 **True** ❽

449 **It removes carbon dioxide** ❶ ❻ ❾

450 **Lithotriptor** ❶ ❸ ❾

589 **Radiotherapy** ❶ ❷ ❾

645 **B: 1903** ❷ ❾

646 **A: 1927** ❺ ❾

648 **C: 1964** ❹ ❾

749 **Radiohead** – released in 1994

845 **Intensive care unit** ❼

848 **Open-heart surgery** ❻ ❾

907 **Cobalt (-60)** – discovered in the late 1930s

908 **Gamma rays, or electron beams** ❷

979 **Respirator, or ventilator** ❺

Core facts ❶

◆ Some machines enable procedures on internal organs, such as the destruction of cancer cells, to be carried out without surgery, or with minimal intrusion. **Lithotriptors**, for example, break up gall, bladder and other stones.

◆ **Heart-lung** and other life-support machines maintain life during heart surgery or when an organ fails. Some devices monitor body function during and after surgery and in emergencies.

◆ Recently introduced nerve stimulation devices can give **pain relief**.

◆ **Radiotherapy**, ultrasound and other radiation machines aid diagnosis and surgery.

◆ Some **implants** are machines (see page 132).

Radiotherapy ❷

X-rays and other forms of high-energy radiation are widely used in radiotherapy to destroy cancerous tumours. Radiation may be generated externally and beamed into the body, often from a **linear accelerator** – a machine that can produce X-rays, gamma rays or electron beams. The dose delivered aims to be exactly right for the type and position of the tumour to be treated. Rapidly growing cancer cells are particularly sensitive to such radiation.

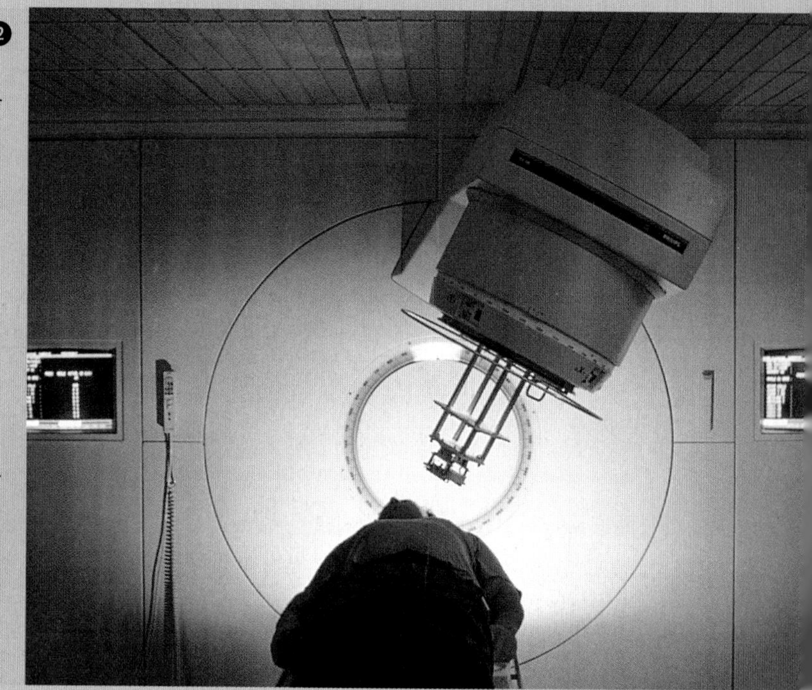

RADIATION TREATMENT The scanner rotates during treatment to minimise damage to healthy tissues surrounding the tumour.

Shock-wave lithotripsy ❸

Painful bladder, gall and kidney stones have traditionally been removed by surgery. A lithotriptor is a modern alternative – a machine that focuses **ultrasonic shock waves** on the stone from outside the body, causing it to shatter into small pieces, which can pass out of the body naturally. X-rays or an ultrasound scan may be used to aim the shock-wave beam accurately. The treatment is uncomfortable, but bruising and discomfort generally ease after a few days.

KIDNEY STONE TREATMENT A lithotriptor focuses sound waves onto the kidneys of a patient immersed in a water bath.

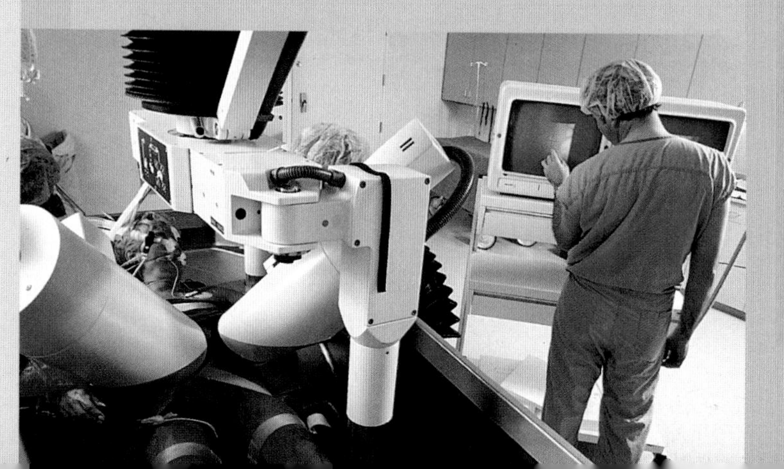

DIY dialysis ❹

A dialysis machine takes over the blood-filtering functions of diseased or failed kidneys. The **home dialysis** machine, developed in the 1960s by US surgeon Belding Scribner, passes fluid into the patient's abdomen and uses the peritoneum (abdomen lining) as a membrane to filter the blood. Now, 21st-century microtechnology enables the production of suitcase-sized, **portable dialysis machines** that can work while patients sleep at night. Wearable and implantable dialysis devices are likely to appear in the near future.

Assisted breathing

A **ventilator** maintains breathing during and immediately after surgery. A tube inserted through the nose into the windpipe carries oxygen, usually from an **oscillating (in-out) pump**. It is much more compact and efficient than the **'iron lung'** that up to the 1950s enabled polio victims to breathe. This enclosed the patient up to the neck and worked indirectly, by applying air pressure to the chest and then releasing it.

Heart-lung machine

A major advance in **open-heart surgery** was the invention of a successful heart-lung machine in the 1950s. This allowed the heart to be stopped, emptied of blood and cooled to prevent deterioration during surgery that may last hours. The machine takes blood from the right atrium (right upper chamber) or venae cavae (main veins into the heart), or from the groin, pumps it through an oxygenating unit (where carbon dioxide is also removed), then returns it to the aorta (main artery) and pumps it around the body.

Monitoring machine

During surgery and in intensive-care units (ICUs), sophisticated electronic devices constantly monitor patients' vital body functions – including heart rate and rhythm, blood pressure and blood oxygen level. They are programmed to raise an alarm if any reading goes above or below normal.

MEASURING FITNESS A woman takes a stress test on an exercise bicycle. Results from heart, blood pressure and respiration monitors are displayed on the screen.

57

Pain relief

Transcutaneous electrical nerve stimulation (**TeNS**) is a high-tech version of the old 'rub it better' cure for pain. Small electric shocks, created by an impulse generator, are passed between two electrodes attached to the skin. The counter-stimulation blocks pain sensations after about 30 minutes. TeNS does not work for everyone, but is often useful for back pain and pain from nerve damage.

ELECTRONIC AID Two small TeNS pads are placed on the wrist to relieve pain.

WEIRD AND WONDERFUL

Dutch doctor Willem Kolff developed the **first kidney machine** in Nazi-occupied Netherlands during World War II. For the dialysis membrane he used sausage skins, but later switched to Cellophane sheets.

TIMESCALE

- ▶ **1901** French physicists Marie and Pierre Curie discover tissue burns caused by radium radiation. Marie Curie wrote a *Treatise on Radioactivity* in 1910.
- ▶ **1903** German surgeon Georg Perthes uses X-ray radiotherapy on a tumour.
- ▶ ***c.*1927** First respirator – the 'iron lung' – made by Harvard medical researcher Philip Drinker.
- ▶ **1943** First kidney machine for human dialysis.
- ▶ **1954** US surgeon John Gibbons performs the first successful open-heart surgery using a heart-lung machine.
- ▶ **1960s** Linear accelerator introduced for radiotherapy.
- ▶ **1964** First use of an automated home kidney dialysis machine.
- ▶ **1980** First shock-wave lithotriptor used to break up kidney stones.

Drug treatment 1

QUESTION NUMBER

The numbers or star following the answers refer to information boxes on the right.

ANSWERS

13	**Hypodermic syringe** (or hypodermic needle) 5
58	**Under the skin** 5
71	**Prescription** – from the Latin 'writing in front'
129	**Quinine** 2
229	**Wormwood** 2
231	**1980s** 7
233	**1960s** – beta blockers slow the heart
234	**1890s** (1899) 7
237	**1800s** (1803) 7
240	**1940s** (1944) 7
258	**D: Gerhard Domagk** (not involved with penicillin) 6
★ 353	**Insulin** 3
356	**A herbal** 2
357	**Willow, or meadowsweet** 2
358	**Opium poppy** 2
359	**Cinchona** 2
360	**Deadly nightshade** 2
383	**True** (in 1632) 2 7
480	**Drugs** 1
669	**Intravenous** 5
759	**An agonist** 3
810	**Pharmacology** 1
938	**True** 4

Core facts 1

◆ From folk and **herbal remedies**, pharmaceuticals (drugs) have become medicine's most important treatments.

◆ **Medicinal drugs** may prevent a disease, stop its progress, or cure a disease altogether, or they may simply relieve symptoms. Some are used for diagnosis rather than treatment.

◆ The past 50 years have seen a burgeoning of drugs, as pharmacologists (chemists) have learned how to extract active ingredients from plant remedies, learn how those ingredients act, and **design new drugs** that are more potent.

◆ It is now often possible to **tailor-make drugs** to have predictable effects on disease.

From herbal beginnings 2

Physicians of old had thousands of herbal remedies that they could use. Those that had real value are still in use today, and pharmacologists search constantly for more plant-based drugs. Examples include:

Artemisinins For centuries, the Chinese used wormwood (Artemisia) to treat malaria. Now drugs derived from it, such as artemether, are among the most powerful antimalarials.

Aspirin The pain-killing effects of willow bark were proven 75 years before the active ingredient, salicin, was isolated. Aspirin was first marketed in 1899.

Atropine This poisonous ingredient of deadly nightshade is used for some heart conditions and to dilate the pupils of the eyes for surgery.

Curare Used by South American Indians as an arrow-tip poison, this extract of bark is a muscle relaxant widely used during surgery.

Digitalis Foxglove leaves were long used as a heart stimulant and the active ingredient, digitalis, is still used today.

Ephedrine Physicians of ancient China and Rome gave asthmatic patients the herb *ma huang* or ephedra – source of the anti-asthma drug ephedrine.

Morphine, heroin and codeine The opiates, derived from the opium poppy, include codeine, a strong but safe painkiller.

Quinine Jesuit missionaries brought the bark of the cinchona tree – source of quinine – from South America to Europe in the 1600s. The people of the Andes used it to treat malaria.

WILLOW BARK An English vicar noted in the 1750s that chewing willow bark relieved his aches and pains.

How drugs work 3

Many drugs attach themselves to **protein molecules** on body cells or on the surface membranes of disease-causing organisms. They reinforce or inhibit the action of **ligands**, natural chemicals that attach themselves to the 'receptor' protein molecules and act as messengers, causing a change or action in the cell or organism. Some drugs, **agonists**, mimic the ligands' action, boosting the 'message'; others, **antagonists**, stop it getting through.

Drug molecule fits receptor site like a key in a lock.

Cell receptor/ligand

Agonist molecule mimics messenger, boosting its effect.

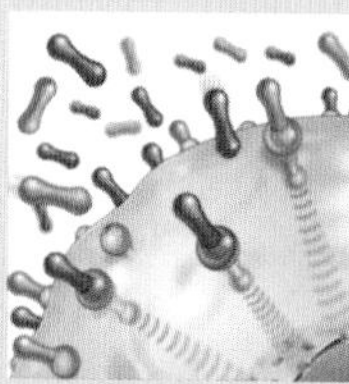

Agonist molecule

Antagonist molecule blocks messenger's access to receptor.

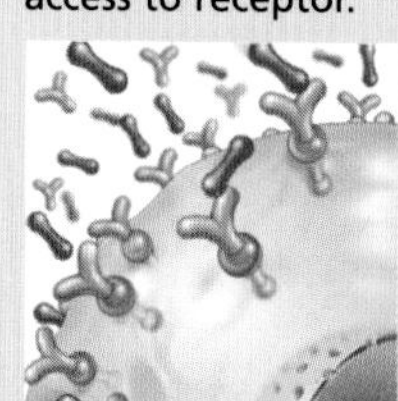

Antagonist molecule

★ 353

Insulin

The hormone insulin, needed by diabetics, was the first drug to be made by **genetic engineering**, which is now used to produce a number of drugs. The human gene for making insulin is spliced into plasmids (bacterial DNA rings), which are then inserted into bacterial cells. When these reproduce they form a large colony of **insulin-manufacturing bacteria**, from which the hormone is extracted.

Designing new drugs

Detailed knowledge of the biochemistry of cells and how the body's chemical messengers work enables **computer modelling** of the receptor sites for drugs, and of potential drugs that might fit them. If pharmacologists know the exact sequence of the amino-acid links in the protein chain of a chemical messenger, they can model the shape of its receptor site on the membrane of a target cell in three dimensions on a computer screen.

In the next step, potential drug molecules that have the right chemical properties to bind with the receptor can be modelled in 3-D, examined on-screen from all angles, and tried for fit with the receptor site. The best-fitting molecules can then be created for real in the laboratory and studied to see if they have the predicted effect.

The search for magic bullets

Twentieth-century pharmacologists searched for 'magic bullets' to destroy specific disease organisms without harming the body. Pioneers were:

Paul Ehrlich (1854-1915) Working on the basis that certain dyes stain some bacteria, Ehrlich looked for substances that would stain and kill them. In 1910 he found salvarsan, an arsenic compound that kills syphilis spirochaetes.

Gerhard Domagk (1895-1964) Domagk discovered the first of the powerful sulpha drugs in 1927, when he found that Prontosil Red dye kills streptococcal bacteria. He announced his finding in 1935.

Alexander Fleming (1881-1955) In 1928 Fleming discovered that a natural mould called penicillium had an 'antibiotic' effect – that is, it killed other bacteria.

Selman Waksman (1888-1973) He isolated the antibiotic streptomycin in 1944.

INVENTING ANTIBIOTICS Biochemist Ernst Chain (right) and pathologist Howard Florey isolated penicillin at Oxford in 1940.

How drugs are given

Herbal remedies were traditionally steeped in water to make infusions or teas, or mixed with water or alcohol to make liquid medicines, called tinctures. Some were powdered and rolled to form pills. These methods are still used, along with high-tech drug-delivery systems.

◆ **Through the skin (transdermal)** Creams and ointments – greasy, fast-acting creams – may still be used for direct treatment. Modern patches have a membrane that allows a controlled amount of a drug to pass through the skin into the blood.

◆ **Implant** Slow-release capsules inserted just under the skin allow a drug to seep into the bloodstream over months.

◆ **Via mucous membrane** A nasal spray or drops or a tablet held in the cheek (buccal) or under the tongue (sublingual) allows the drug to diffuse through the mucous membrane lining of the nose or mouth. It ensures faster absorption than through the skin, at a steady rate.

◆ **Suppositories and enemas** A drug in a suppository, enema or foam inserted into the rectum or vagina is absorbed quickly into the blood. Vaginal pessaries are normally used to treat local gynaecological disorders.

◆ **Inhalation** An inhaler (puffer) produces a fine mist containing a drug that is breathed in and acts almost instantaneously on the lungs. It is commonly used for asthma treatments.

◆ **Oral** The route for pills, liquid medicines and capsules; the drug is absorbed from the stomach or intestines. Slow-release capsules contain tiny balls of the drug that dissolve at different speeds.

◆ **Injection** Hypodermic syringes may be used to inject a drug into the skin (percutaneous), under it (subcutaneous), or into a muscle (intramuscular) or a vein (intravenous). Injection guns use compressed gas to 'shoot' a drug painlessly into the skin or another membrane.

◆ **Intravenous injection/infusion** A drug injected directly into a vein takes effect rapidly.

◆ **Drug-delivery pump** A tiny electronic device, either with a timer or under manual control, can be implanted under the skin to deliver regular medication over a long period.

AMPOULES Sealed containers for medicines ensure that the drug remains sterile until use.

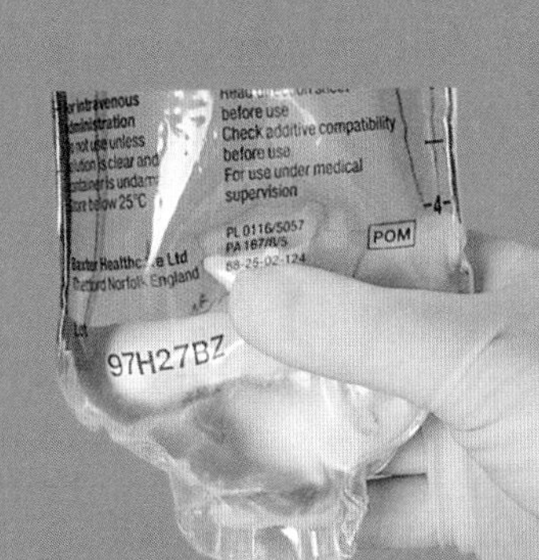

INTRAVENOUS DRIP This device infuses a drug slowly into a vein through a cannula (narrow tube) and a needle.

TIMESCALE

▶ **1632** Quinine is introduced to Europe.
▶ **1785** Digitalis used as a heart stimulant.
▶ **1803** Morphine is extracted from opium.
▶ **1840s** First anaesthetic drugs in use.
▶ **1853** Hypodermic syringe invented.
▶ **1899** Aspirin produced from willow bark.
▶ **1902** Veronal, the first barbiturate, in use.
▶ **1910** Salvarsan produced to cure syphilis.
▶ **1928** Penicillin found.
▶ **1927-35** Sulpha drug from Prontosil Red dye.
▶ **1940** Penicillin is isolated.
▶ **1940s-50s** Sex hormones synthesised.
▶ **1944** Streptomycin discovered.
▶ **1949** Lithium used to treat bipolar disorder.
▶ **1950** Chemotherapy used to treat leukaemia.
▶ **1959** Benzodiazepine tranquillisers produced.
▶ **1971** Tamoxifen used to treat breast cancer.
▶ **1980s** Antiviral drugs.
▶ **1987** Fluoxetine released as Prozac.
▶ **2000s** Statins known to prevent heart disease.

Drug treatment 2

QUESTION NUMBER

The numbers or star following the answers refer to information boxes on the right.

ANSWERS

59	**Applied to the skin (as a cream, etc)** ❽
60	**Hypertension, or high blood pressure** ❹
121	**Adrenaline** ❼
124	**Aspirin** ❸
126	**Ranitidine ('Zantac')** ❺
127	**Digoxin** ❹
160	**Cyclosporin** (an immunosuppressant) ❼
202	**Malaria** ❷
208	**Pass more urine** ❹
210	**Histamine** ❼
347	**Constipation** ❺
355	**Breast cancer** ❽
573	**Syringe** – hypodermic invented in France, in 1850s
★ 594	**On a prescription**
630	**Anticoagulants, or anti-blood-clotting** ❹
655	**The lungs, or bronchi** ❽
708	**Sleeping tablets** ❸
709	**Thrombolytics** ❹
710	**A pharmacopeia, or formulary** ❶
760	**Antiviral** ❷
900	**Antihistamine** ❼
909	**Poisons or kills cells** ❽

List of major drugs ❶

Books listing drugs available to doctors to prescribe – known as pharmacopeias and formularies – include thousands of substances used to treat disease, but the substances fall into a much smaller number of categories. Important groups and types of drugs include:

Anti-infective drugs ❷

◆ **Antibiotics; other antibacterials** Strictly, an antibiotic is an antibacterial drug made by another organism such as a bacterium or mould, or a related substance. But sulpha drugs and other antibacterials may also be called antibiotics. True antibiotics include the penicillins, tetracyclines and aminoglycosides (streptomycin, for example). In a special category are those used to treat tuberculosis (TB), which is resistant to many antibiotics.

◆ **Antiviral drugs** These generally interfere with the reproduction of viruses inside body cells, or stop them entering those cells – and they should do this without harming the body cells. Examples include anti-HIV drugs such as zidovudine (AZT).

◆ **Antiprotozoal drugs** These act against protozoal infections, including malaria. There are no antimalarial vaccines, so in parts of the world where the disease is endemic, people need to take regular antimalarial medication. Quinine and arteminisins have been used against malaria for centuries.

◆ **Others** Antifungal drugs (against thrush, fungal skin infections and others) and anthelmintics (to eradicate parasitic worms).

Acting on the nervous system ❸

◆ **Painkillers** Opiate painkillers (from opium) such as morphine and codeine act on the brain, blocking the transmission of pain impulses. Non-opiates such as aspirin and paracetamol are less powerful but safer. Paracetamol blocks pain signals in the brain. Aspirin and ibuprofen are non-steroidal anti-inflammatory drugs (NSAIDs), which act by blocking pain and prostaglandins that trigger inflammation.

◆ **Antimigraine drugs** This class of drugs act on blood vessels in the brain. They prevent or correct the constriction followed by widening of blood vessels that takes place during migraine attacks.

◆ **Anaesthetics** These block all feeling, not just pain, and are used before surgery.

◆ **Anticonvulsants** Drugs in this group reduce the high electrical activity in the brain that causes epileptic seizures. They include some barbiturates, tranquillisers such as diazepam ('Valium') and specialised drugs.

◆ **Hypnotics** These drugs help to correct insomnia. Barbiturates such as phenobarbitone were the main hypnotics, but these are dangerous in an overdose. The main ones used today are short-acting benzodiazepine tranquillisers such as temazepam, and other drugs that act similarly on brain receptors to reduce brain activity. Some antihistamines (see box 7) are used as mild hypnotics.

◆ **Psychoactive drugs** Used to treat mental disorders (see page 100).

◆ **Stimulants** Amphetamines and others may be used in the sleep disorder narcolepsy and for attention deficit disorder (ADD).

Drugs for blood disorders ❹

◆ **Drugs for hypertension** Drugs to lower blood pressure also prevent heart disease and strokes. They are often used in combination. The main types include angiotensin-converting enzyme (ACE) inhibitors, which act on enzymes controlling the constriction of blood vessels; beta blockers, which counteract adrenaline and noradrenaline, hormones affecting the heart rate and blood pressure; calcium-channel blockers, which cause blood vessels to relax; and diuretics.

◆ **Diuretics** 'Water pills' increase the amount of urine passed by the kidneys. They are used to treat high blood pressure, but also chronic heart failure and other conditions in which fluid tends to accumulate in the tissues – including some liver and kidney disorders, and altitude sickness.

◆ **Drugs for heart conditions** These fall into three main groups: anti-arrhythmics, such as beta blockers, which help regulate an uneven or a too slow or fast heartbeat; heart stimulants, mainly digitalis and digoxin, which strengthen the heart's action; and nitrates, used to treat angina.

◆ **Drugs for clotting disorders** Haemophilia may be corrected with vitamin K, blood products such as Factor VIII, or other drugs. Warfarin and heparin are anticoagulants used to correct blood that is abnormally prone to form clots. Thrombolytic drugs (or 'clot-busters') dissolve blood clots and are used only as emergency treatment for heart attacks and strokes.

◆ **Lipid-lowering drugs** Statins and related drugs reduce the amount of lipids (cholesterol and other fatty substances) in the blood to reduce the risk of heart attack.

Digestion and nutrition 5

◆ **Antiemetics** These prevent or relieve nausea and vomiting caused by travel sickness, migraine, inner ear disorders and chemotherapy for cancer. They suppress nerve signals to or from the brain centre that controls vomiting.
◆ **Antacids and other drugs** These reduce excess stomach acid to relieve heartburn and indigestion. Antacids such as aluminium hydroxide and sodium bicarbonate neutralise the acid. Proton-pump inhibitors and H2 blockers such as ranitidine (Zantac) reduce the amount of acid produced.
◆ **Antidiarrhoeals and laxatives** These relieve diarrhoea and constipation respectively. Some act by reducing or stimulating intestinal contractions. Longer-term, bulk-forming materials such as bran can correct both conditions. Osmotic laxatives such as lactulose retain water in the faeces, softening them.
◆ **Dietary supplements** Vitamin and/or mineral supplements are valuable if the diet is deficient in the natural substances.

Hormones and related drugs 6

◆ **Corticosteroids** Drugs that are similar to cortisone and other hormones produced by the adrenal glands help the body to cope with stress. They may be given as creams and ointments (topical corticosteroids) to treat some skin conditions; by direct injection for conditions such as rheumatoid arthritis and tennis elbow; or by mouth for severe rheumatoid arthritis or asthma. Some corticosteroids are also used as immunosuppressants after transplants.
◆ **Female hormones** Oestrogen and/or progestogen are used for contraception and hormone-replacement therapy. Other hormones and drugs are used for fertility treatment, for inducing labour, and for other conditions during or after childbirth.
◆ **Male hormones** Anti-androgen drugs are used to treat prostate gland problems.
◆ **Others** Insulin and other drugs to treat diabetes; thyroxine derivatives to treat an underactive thyroid; drugs to treat an overactive thyroid; drugs such as growth hormone inhibitors, that affect the pituitary.

Acting on the immune system 7

◆ **Anti-allergy drugs** These counter allergic reactions such as atopic eczema and hay fever. They include antihistamines, which block the effects of the histamine released in allergic reactions; some cause drowsiness and may also be used for sleep disturbances. The hormone adrenaline is used to counter anaphylactic shock, a severe allergy.
◆ **Immunosuppressants** Drugs to suppress the immune response. They are used to treat autoimmune diseases and to counter the rejection of transplants. They include corticosteroids, cytotoxic drugs (which stop white blood cells growing), and others such as the antibiotic cyclosporin.

Other drugs 8

◆ **For skin conditions** These include antiseptics and antibiotics to treat skin infections; various antipruritics, which reduce itching; topical corticosteroids of various strengths to treat inflammatory and itchy conditions such as eczema; retinoid drugs used to treat acne and psoriasis; and emollient (moisturising) creams for dry skin.
◆ **For joint problems** Corticosteroids are sometimes used, especially for arthritis, but less in tablet form than in the past because of side-effects. Instead, non-steroidal anti-inflammatory drugs (NSAIDs) such as ibuprofen, diclofenac and others (including aspirin, which is weakly anti-inflammatory) are usually prescribed. Severe rheumatoid arthritis may be treated with antirheumatic drugs, including immunosuppressants.
◆ **For respiratory problems** This category includes cough remedies, decongestants (which reduce swelling of the membranes in the nose and sinuses), and bronchodilators (which act at a lower level, in the lungs, for asthma and chronic bronchitis, for example).
◆ **Against cancer** Cytotoxic drugs, which kill or damage cancerous and non-cancerous cells, are the main ones used. Their effect is greatest on cells like cancer cells that are rapidly growing, but their effect on ordinary cells means that there has to be a period of recovery between treatments. Other anticancer drugs include hormones and hormone antagonists, which slow the growth of some tumours – sometimes enough for the body's defences to destroy the cancer. Such drugs include tamoxifen and anastrazole for breast cancer, and buserelin and goserelin for prostate cancer.

BACKGROUND IMAGE Hospital cancer ward pharmacists wear protective clothing and work in glove boxes when preparing drugs for chemotherapy.

★ 594

Old measures

Until January 1971, UK pharmacists often weighed and measured drugs in the **medieval Apothecaries' system**: weights in grains, scruples, drachms and Troy ounces; volumes in minims and fluid drachms.

1 grain	= 0.0648 grams
1 scruple	= 20 grains
1 drachm	= 3 scruples
1 Troy ounce	= 8 drachms (31.1grams)
1 minim	= 0.059 ml (vol. of 1 grain of water)
1 fl drachm	= 60 minims (3.55 ml)

QUESTION NUMBER

The numbers or star following the answers refer to information boxes on the right.

ANSWERS

- **79** **Designer babies** – may involve embryo selection
- **118** ***Fantastic Voyage*** – 1966, directed by Richard Fleischer
- **119** **Nanotechnology** 9
- **120** **Dendrimers** 9
- **179** **Genetic screening** 2
- **180** **Monoclonal antibodies** 6
- **248** **Stem cells** 4
- ★ **352** **Smart bombs**
- **399** **A virus** 3
- **400** **Hybridoma** 6
- **484** **Michelle Whitaker** – baby Jamie born in Sheffield
- **499** **True** 4
- **910** **The organ comes from a different species** 7
- **933** **True** 8

Frontiers of medicine

Core facts

- The focus of medical science at the start of the 21st century is on finding cures for **degenerative**, **functional**, **inherited** and **malignant** diseases.
- Discoveries, particularly in **gene therapy**, **cell cloning** and **stem cell** techniques, are leading to advances in treating these disorders.
- New ways must also be found to fight diseases caused by **antibiotic-resistant bacteria**, and new **viral diseases** such as SARS.
- **Nanotechnology** (microscopic engineering) is expected to bring advances in treatment.
- Many new **genetic techniques** are controversial and under tight legal control.

Genetic screening 2

Genetic screening of children and adults is well established. It is possible to test for an **inherited disorder** that may arise in later life, or to give counselling on the risk of children inheriting a **genetic disorder**.

Some techniques are controversial. These include the screening of cells taken from embryos made by in-vitro fertilisation (IVF) to check for abnormalities before they are implanted in a woman's womb; and genetic matching – checking all embryos in an IVF 'batch' to find one that is closest genetically to a sibling who may in future need a bone-marrow or other transplant.

Gene therapy 3

In theory, the defective genes of people with inherited disorders could be **replaced with normal genes** using genetic engineering methods. One method, successful in some animal experiments, uses a virus as a 'vector', or carrier, to insert the replacement gene in the recipient's DNA. Problems with this technique include the risk of the vector virus harming the recipient, or the recipient's immune system destroying the virus. If the therapeutic DNA is not established in stable cells that reproduce, the benefits are short-lived. A few human trials have already taken place.

MAPPING THE HUMAN GENOME The background image shows part of the human gene map, which is playing an important role in the diagnosing and treatment of diseases.

Stem cells 4

These are the body's **'master cells'**. They are human cells that have not yet specialised as, for example, liver, brain or muscle cells, so they can grow into any type of tissue with the appropriate chemical or other trigger. Research suggests that stem cells could be used to cure disorders such as Parkinson's disease and diabetes if injected into the body. Controversially, the main source of stem cells is aborted foetuses or unused IVF embryos.

HUMAN STEM CELLS These cells could one day be used to 'grow' entire body organs for implantation.

Cloning 5

This is the **creation of genetically identical cells or organisms** by inserting an individual's genetic material – DNA – in place of the DNA removed from an egg of the same species. There have been claims of human clones, but ethical concerns have led to many countries banning human cloning. However, some scientists believe that the cloning of human non-reproductive cells is essential for research into new methods of treatment.

CREATING A CLONE A pipette holds a sheep's egg from which the nucleus has been removed. The cell in the needle on the right is about to be injected into the egg.

Antibodies against antigens

6

Special cells, known as **monoclonal antibodies**, are created in laboratories to diagnose certain diseases and to identify tumour cells by binding with them. Antibodies are proteins in the body that attach themselves to the 'foreign' antigens of disease organisms such as viruses, and neutralise them. Monoclonal antibodies differ from our antibodies in that they are all identical and act against one antigen. They are made by uniting an antibody-producing white blood cell (a B-cell) with a tumour cell to make a hybrid called a hybridoma. This cell will grow indefinitely (forming clones), and continuously supply the B-cell's specific antibody.

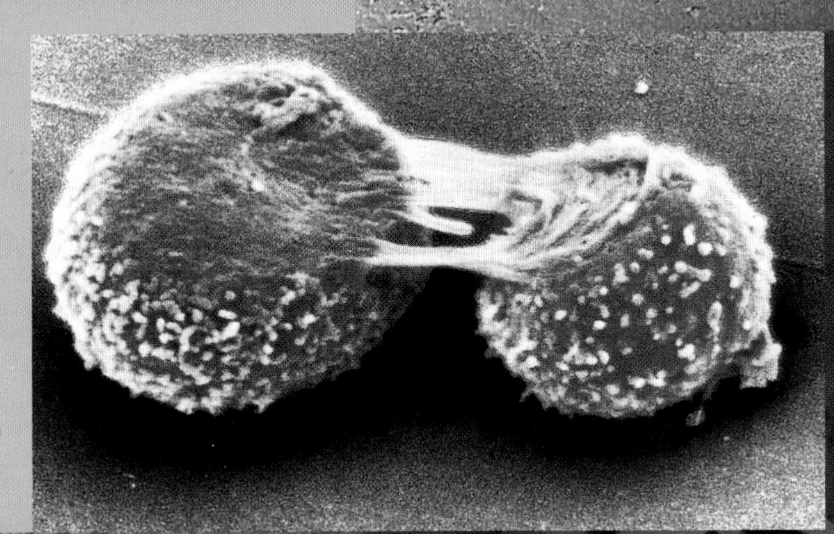

RESEARCH TOOL These hybridoma cells are being used in experiments to deliver cell-destroying drugs to cancer cells.

Transgenics

7

Pigs' organs are very similar in size and function to human organs – **pig heart valves** are widely used in heart surgery. Although the human immune system rejects whole pig's organs, pig's **livers**, for example, have been used to keep human patients alive for a short period until a human donor organ is found. Researchers have been working on rearing genetically engineered animals whose tissues will not be rejected, but there is a risk that the animal organ may introduce possibly unknown viral diseases.

WEIRD AND WONDERFUL

8

Cornflakes made from maize genetically engineered to produce medicines are a possible future development. A GM tobacco grown today contains a genital herpes treatment, and experimental crops have produced a blood-clotting agent.

352

Viral bombs

Therapeutic 'smart bombs' can target and destroy **disease cells** with the accuracy of military bombs. A treatment for the brain cancer, malignant glioma, will use a virus engineered to reproduce in and kill cancer cells only. Alternatively, a drug could be delivered to the core of a tumour cell. This might be done by combining an anticancer drug with a protein-nucleic acid chain that will bind to the RNA in the nucleii of the cancer cells forming the tumour.

NANOROBOT A micro-submarine like this could unblock an artery from within.

Nanotechnology

9

Medical micromachines are tiny devices that can be implanted to monitor body functions. **Minute microelectronic-controlled robots** have been used to perform surgery. But true nanomachines would be comparable in size to large molecules. **Dendrimers**, a type of polymer discovered in the 1980s, could be components of such machines. They have many intricately branching arms and can be built to order. A number of dendrimer molecules could be combined to make molecular machines with genetic parts to identify, say, cancer cells. Such a nanomachine could be injected into the body to look for cancer cells, deliver a drug to kill them, then check the results.

QUESTION NUMBER

The numbers or star following the answers refer to information boxes on the right.

ANSWERS

Question	Answer
3	**Autopsy** ❹
11	**After death** ❹
19	**Rigor mortis** – 'the stiffness of death'
197	**True** ❶
201	**HIV/AIDS** ❸ ❺
230	**(c); (d); (a); (b); (e)** ❻
388	**False** (it was 30 times higher) ❶
410	**D: 27 000** ❸
430	**Psalms** (Psalm 90, verse 10) ❼
493	**False** (about 110) ❼
590	**The brain** ❶ ❷
★ 603	**Cryonics**
660	**Hair and nails** – for several hours after death
670	**Post mortem** ❹
939	**False** – less than half, despite many more cars

Death and its causes

Core facts

◆ 'Brain death', the cessation of brain function, is the accepted way of **defining death** today.
◆ In Britain and worldwide, the major **cause of death** is heart and circulatory disease.
◆ **Patterns of mortality** change over time. In recent years, deaths from heart disease have declined, but not those from cancer.
◆ Infant and **childhood mortality** are still significant, especially in the developing world. In advanced countries such as the UK infant deaths in 1900 were 30 times higher than today.
◆ **Life expectancy** has generally increased. Today 80 per cent of people in Britain live over 60 years, compared with 20 per cent in 1900.

Defining death ❷

In the past, a person was considered dead if the heart and breathing stopped, but stopping and restarting these are now routine in surgery. However, if the brain stops receiving oxygen from circulating blood for more than a very few minutes its cells start to die. After a further short period, it might be possible to revive the person but at the expense of irreversible brain damage, because the parts dealing with higher thought processes die first. A little longer, and the brain cannot recover, because the brainstem, which controls basic bodily functions, ceases to work. This is **brain death** – the point at which most doctors and lawyers consider life to have ended.

Patterns of mortality ❸

Infectious diseases used to be the main killers. During the 1300s, the plague killed an estimated third of Europe's population. As late as the early 20th century, tuberculosis, diphtheria, flu, measles and other infections killed huge numbers of people. But with vaccination and better treatment, infections declined as a cause of death until the HIV/AIDS epidemic emerged in the 1980s.

Heart and other circulatory diseases and **cancers** became the main causes of dying during the 20th century, but heart disease is significantly less of a killer now than in 1950. Infant mortality has also fallen dramatically from 27 000 deaths of children under one year old in 1949 to 2500 in 1995.

Post mortem ❹

An **autopsy** (also called a necropsy or post mortem – 'after death') is a medical examination of a dead body to determine the cause of death. It is usually carried out by a **pathologist** (a medical scientist specialising in disease and its effects) if the cause of death is not clear, or for other medical or legal reasons.

The pathologist examines the body externally and usually cuts it open to examine and possibly remove internal organs. It may also be necessary to examine small samples of body tissues under a microscope for tell-tale changes caused by disease. Biochemical or other tests may be carried out to detect poisoning and other possible unnatural causes of death.

BACKGROUND IMAGE
A pathologist at work in the autopsy room of a modern hospital.

Leading causes of death

5

The pie chart below represents the major groups of diseases and disorders that caused deaths worldwide in 2002. The table lists the diseases in each group that caused 1 per cent or more deaths in England and Wales compared with the rest of the world. The gaps are where a disease caused less than 1 per cent of deaths. The table is based on World Health Organisation (WHO) figures.

PERCENTAGES OF ALL DEATHS

(2002 – percentages of all deaths unless indicated)

Disease	Worldwide	England and Wales
Heart and circulatory disorders	**29.2%**	**39.3%**
Coronary heart disease	12.6%	27.8%
Stroke	9.6%	6.3%
Other heart disease	2.9%	
Infectious diseases	**19.5%**	
HIV/AIDS	4.9%	
Diarrhoeal diseases	3.1%	
Tuberculosis	2.8%	
Malaria	2.1%	
Measles	1.3%	
Cancer	**12.5%**	**26.3%**
Lungs and trachea	2.2%	5.4%
Breast	*1.8%	*4.1%
Stomach	1.5%	1.1%
Colorectal	1.1%	2.7%
Liver	1.1%	
Cervix	*0.9%	
Prostate	**0.9%	**3.5%
Oesophagus	0.8%	1.2%
Ovary	*0.5%	*1.5%
Accidents and injuries	**9.1%**	**3.0%**
Respiratory disorders	**13.2%**	**13.1%**
Chronic bronchitis and emphysema	4.8%	4.5%
Pneumonia	6.7%	6.1%
Perinatal conditions (at or near birth)	**4.3%**	
Digestive diseases	**3.4%**	**4.5%**
Cirrhosis of the liver	1.4%	1.2%
Psychiatric and nervous system disorders	**1.9%**	
Diabetes	**1.7%**	**1.2%**
Urinary and genital diseases	**1.5%**	**1.6%**
Kidney diseases	1.2%	
Others	**3.7%**	

* % of women only. ** % of men only. All others: % of all deaths.

Life expectancy at birth

6

Top 12 countries, 2001

Country	Age
Andorra	83.5
Japan	81.6
Sweden	80.1
Hong Kong	79.9
Iceland	79.8
Canada	79.3
Spain	79.3
Australia	79.2
Israel	79.2
Martinique	79.1
Switzerland	79.1
France	79.0

Bottom 12 countries, 2001

Country	Age
Burundi	40.9
Angola	40.1
Botswana	39.7
Cen. African Rep.	39.5
Rwanda	39.3
Mozambique	38.1
Malawi	37.5
Lesotho	35.1
Swaziland	34.4
Sierra Leone	34.2
Zimbabwe	33.1
Zambia	32.4

Germany 78.3 (21st); **UK** 78.2 (27th); **USA** 77.1 (34th)

Limits of the lifespan

7

Most people have the potential to exceed comfortably the biblical lifespan of 'threescore years and ten'. But geneticists believe that most of the body's cells are **programmed to die** after a certain number of divisions. As a result, short of extensive genetic engineering, there is a natural limit to how long a person can live, because the organs inevitably deteriorate to the point where they can no longer sustain the life of the whole person. Although there are exceptions – Frenchwoman **Jeanne Louise Calmet**, for example, lived to the ripe old age of **122 years** – the natural limit for most people is believed to be around 110 years. Diseases, disorders and accidents ensure that a large majority of people die well short of that.

Cryonics

Since the 1960s, **deep-freezing** a person at the moment of death in the hope of **future resuscitation** has been a growing business, especially in the USA. But most doctors have serious doubts about the possibility of thawing out a corpse and curing whatever caused death. Even if freezing does not damage the dead person's tissues, the brain cells and other tissues are likely to have decayed beyond repair in the time between death and freezing.

Nobel prize winners

QUESTION NUMBER

The numbers or star following the answers refer to information boxes on the right.

ANSWERS

249 Karolinska Institute ❶

250 Argentinian ❶

331 Wilhelm Roentgen ❸

332 Robert Koch ❶

333 Willem Einthoven ❶

334 Thomas Hunt Morgan ❶

335 Gerhard Domagk ❶

336 Selman Waksman ❶

337 Frederick Sanger ❷

338 Godfrey Houndsfield ❶

339 Kary Mullis ❷

340 Stanley Prusiner ❶

371 Frederick Sanger ❷

435 **1930** (accept 1927-33) ❶

★ 446 MRI scanning

691 **Nobel prize ceremony** – directed by Mark Robson

820 **Rosalind Franklin** – the first to take X-rays of DNA

886 Dr Gertrude Elion ❶

980 Diphtheria ❶

WINNER IN 1901 German bacteriologist Emil von Behring.

Physiology or medicine ❶

1901 **Emil Adolf von Behring** Diphtheria antitoxin
1902 **Ronald Ross** Malaria
1903 **Niels Ryberg Finsen** Light radiation therapy
1904 **Ivan Pavlov** Physiology of digestion
1905 **Robert Koch** Cause and treatment of tuberculosis
1906 **Camillo Golgi and Santiago Ramón y Cajal** Structure of nervous system
1907 **Alphonse Laveran** Protozoal diseases
1908 **Paul Ehrlich and Ilya Mechnikov** Immunity
1909 **Theodor Kocher** Function, disease and surgery of the thyroid
1910 **Albrecht Kossel** Cell chemistry
1911 **Allvar Gullstrand** Optics of the eye
1912 **Alexis Carrel** Blood-vessel surgery
1913 **Charles Richet** Allergies
1914 **Robert Bárány** Balance function of inner ear
1915 No prize awarded
1916 No prize awarded
1917 No prize awarded
1918 No prize awarded
1919 **Jules Bordet** Immunity
1920 **August Krogh** Action of capillaries
1921 No prize awarded
1922 **Archibald Hill; Otto Meyerhof** Muscles
1923 **Frederick Banting and John Macleod** Insulin
1924 **Willem Einthoven** Electrocardiogram
1925 No prize awarded
1926 **Johannes Fibiger** Cancer-causing parasite
1927 **Julius Wagner-Jauregg** Effects of fever on dementia
1928 **Charles Nicolle** Typhus
1929 **Christiaan Eijkman; Sir Frederick Gowland Hopkins** Discovery of vitamins
1930 **Karl Landsteiner** Discovery of blood groups
1931 **Otto Warburg** Respiratory enzymes
1932 **Edgar Adrian and Charles Sherrington** Neuron function
1933 **Thomas Hunt Morgan** Role of chromosomes
1934 **George Minot, William Murphy and George Whipple** Liver treatment for anaemia
1935 **Hans Spemann** Embryonic development
1936 **Sir Henry Dale and Otto Loewi** Nerve impulses
1937 **Albert Szent-Györgyi** Cellular combustion
1938 **Corneille Heymans** Regulation of respiration
1939 **Gerhard Domagk** Prontosil (first sulpha drug)
1940 No prize awarded
1941 No prize awarded
1942 No prize awarded
1943 **Henrik Dam; Edward Doisy** Vitamin K
1944 **Joseph Erlanger and Herbert Gasser** Nerve fibres
1945 **Sir Alexander Fleming, Ernst Chain and Sir Howard Florey** Discovery and production of penicillin
1946 **Hermann Muller** Mutations caused by X-rays
1947 **Carl and Gerty Cori; Bernardo Houssay** Insulin and sugar metabolism
1948 **Paul Müller** DDT's role as insecticide
1949 **Walter Hess; Egas Moniz** Brain organisation; use of prefrontal leucotomy
1950 **Philip Hench, Edward Kendall and Tadeus Reichstein** Adrenal cortex hormones
1951 **Max Theiler** Yellow fever vaccine
1952 **Selman Waksman** Discovery of streptomycin
1953 **Hans Krebs; Fritz Lipmann** Mechanism and enzymes of cellular respiration
1954 **John Enders, Frederick Robbins and Thomas Weller** Culture of polio virus
1955 **Hugo Theorell** Oxidation enzymes
1956 **André Cournand, Werner Forssmann and Dickinson Richards** Cardiac catheterisation
1957 **Daniel Bovet** Antihistamine drugs
1958 **George Beadle and Edward Tatum; Joshua Lederberg** Gene action and organisation
1959 **Arthur Kornberg and Severo Ochoa** Synthesis of nucleic acids
1960 **Sir Macfarlane Burnet and Peter Medawar** Acquired immunotolerance (transplants)

WINNER IN 1930 Austrian doctor Karl Landsteiner won for identifying the ABO blood groups.

1961 **Georg von Békésy** Mechanism of hearing
1962 **Francis Crick, James Watson and Maurice Wilkins** Structure and function of DNA
1963 **Sir John Eccles, Alan Hodgkin and Andrew Huxley** Excitation and inhibition of nerve impulses
1964 **Konrad Bloch and Feodor Lynen** Cholesterol
1965 **François Jacob, André Lwoff and Jacques Monod** Genetic control of enzyme and viral synthesis
1966 **Charles Huggins; Peyton Rous** Treatment of prostate cancer; tumour viruses

1967 **Ragnar Granit, H. Keffer Hartline and George Wald** Eye physiology and chemistry
1968 **Robert Holley, Gobind Khorana and Marshall Nirenberg** How genes act
1969 **Max Delbrück, Alfred Hershey and Salvador Luria** Bacteriophage viruses
1970 **Julius Axelrod, Ulf von Euler and Sir Bernard Katz** Neurotransmitter molecules
1971 **Earl Sutherland, Jr** How hormones act
1972 **Gerald Edelman and Rodney Porter** Chemical structure of antibodies
1973 **Karl von Frisch, Konrad Lorenz and Nikolaas Tinbergen** Animal behaviour
1974 **Albert Claude, Christian de Duve and George Palade** Cell structure and organisation
1975 **David Baltimore, Renato Dulbecco and Howard Temin** Action of tumour viruses on genes of cells
1976 **Baruch Blumberg and Carleton Gajdusek** New mechanisms for origin and spread of infections
1977 **Roger Guillemin and Andrew Schally; Rosalyn Yalow** Hormone research
1978 **Werner Arber, Daniel Nathans and Hamilton Smith** Role of restriction enzymes in molecular genetics
1979 **Allan Cormack and Godfrey Hounsfield** Development of CAT scanner
1980 **Baruj Benacerraf, Jean Dausset and George Snell** Genetic regulation of immunity
1981 **Roger Sperry; David Hubel and Torsten Wiesel** Brain organisation; brain processing of visual data
1982 **Sune Bergström, Bengt Samuelsson and John Vane** Prostaglandins, related substances
1983 **Barbara McClintock** Discovery of 'mobile genes' which change position on chromosomes
1984 **Niels Jerne, Georges Köhler and César Milstein** Control of immune system and monoclonal antibody production
1985 **Michael Brown and Joseph Goldstein** High blood cholesterol and heart disease
1986 **Stanley Cohen and Rita Levi-Montalcini** Discovery of cellular growth factors
1987 **Susumu Tonegawa** Genetic control of antibody production
1988 **Sir James Black, Gertrude Elion and George Hitchings** Action of drugs on cellular receptors; anticancer drugs
1989 **Michael Bishop and Harold Varmus** Role of oncogenes in causing cancer
1990 **Joseph Murray and Donnall Thomas** Organ and bone-marrow transplants
1991 **Erwin Neher and Bert Sakmann** Single-ion channels in cell membranes
1992 **Edmond Fischer and Edwin Krebs** Phosphorylation of proteins (important mechanism of cell regulation)
1993 **Richard Roberts and Phillip Sharp** 'Split genes'
1994 **Alfred Gilman and Martin Rodbell** Discovery of 'G-proteins', cellular signals
1995 **Edward Lewis, Christiane Nüsslein-Volhard and Eric Wieschaus** Genetic control of embryo development
1996 **Peter Doherty and Rolf Zinkernagel** Immune system recognition of virus-infected cells
1997 **Stanley Prusiner** Prions as cause of disease
1998 **Robert Furchgott, Louis Ignarro and Ferid Murad** Role of nitric oxide in blood vessels
1999 **Günter Blobel** Protein transport within cells
2000 **Arvid Carlsson, Paul Greengard and Eric Kandel** Signal transmission in brain
2001 **Leland Hartwell, Tim Hunt and Sir Paul Nurse** Regulation of cells' cycle of growth and division
2002 **Sydney Brenner, Robert Horvitz and John Sulston** How genes regulate organ development and programmed cell death
2003 **Sir Peter Mansfield and Paul Lauterbur** Magnetic resonance imaging (MRI)

Chemistry ❷

When awarded for discoveries significant in medicine.

1902 **Emil Fischer** Sugar and purine synthesis
1907 **Eduard Buchner** Biochemical researches
1928 **Adolf Windaus** Sterols and vitamins
1930 **Hans Fischer** The colouring matter of blood
1937 **Norman Haworth; Paul Karrer** Vitamins
1938 **Richard Kuhn** Carotenoids and vitamins
1939 **Adolf Butenandt** (shared) Sex hormones
1946 **James Sumner**; **John Northrop and Wendell Stanley** Enzymes and virus proteins
1948 **Arne Tiselius** Nature of serum proteins
1955 **Vincent du Vigneaud** Synthesis of hormone
1957 **Lord Todd** Cell proteins
1958 **Frederick Sanger** Structure of insulin molecule
1962 **John Kendrew; Max Perutz** Globular proteins
1964 **Dorothy Hodgkin** X-ray studies of vitamins
1970 **Luis Leloir** Carbohydrate metabolism
1972 **Christian Anfinsen; Stanford Moore and William Stein** Enzyme chemistry
1978 **Peter Mitchell** Cellular energy transfer
1980 **Paul Berg; Walter Gilbert and Frederick Sanger** Chemistry of nucleic acids

WINNERS IN 2003 Sir Peter Mansfield (top) of the UK and Paul Lauterbur (above) of the USA.

1982 **Aaron Klug** Nucleic acid-protein complexes
1984 **Bruce Merrifield** Peptide synthesis
1985 **Herbert Hauptman and Jerome Karle** Biological molecules' crystal structures
1989 **Sidney Altman and Thomas Cech** Catalytic properties of RNA
1993 **Kary Mullis; Michael Smith** DNA amplification; genetic engineering techniques
1997 **Paul Boyer and John Walker; Jens Skou** ATP chemistry and function
2002 **John Fenn and Koichi Tanaka; Kurt Wüthrich** Methods of identifying and analysing biological macromolecules
2003 **Peter Agre; Roderick MacKinnon** Channels in cell membranes

Physics ❸

When awarded for discoveries significant in medicine.

1901 **Wilhelm Roentgen** X-rays
1915 **Lawrence and William Bragg** X-ray crystallography
1952 **Felix Bloch and Edward Mills Purcell** Nuclear magnetic resonance (basis of MRI)
1953 **Frits Zernike** Phase-contrast microscope (widely used in medicine)
1986 **Ernst Ruska; Gerd Binnig and Heinrich Rohrer** Electron microscopy

★ 446

A long wait

Paul Lauterbur and **Peter Mansfield** won the 2003 Nobel prize for medicine for developing MRI scanning. Recognition had taken decades. The discoverers of magnetic resonance won the 1952 prize for physics, and Lauterbur and Mansfield began to apply it to body imaging in the 1970s. When they won the prize, over 20 000 MRI scanners were in use worldwide.

Quiz 0 (page 8)

1 Apothecary
2 Anorexia (nervosa)
3 Autopsy
4 Anaesthetic
5 Acupuncture
6 Aromatherapy
7 Adrenaline
8 Arteries
9 Allergy
10 Amputation

Quiz 1 (page 8)

11 After death
12 Local anaesthetic
13 Hypodermic syringe
14 Tourniquet
15 Epilepsy
16 Jaundice
17 Chiropractic
18 Keyhole surgery
19 Rigor mortis
20 Foetus

Quiz 2 (page 8)

21 True
22 True
23 True
24 False
25 True
26 True
27 False
28 False
29 True
30 True

Quiz 3 (page 8)

31 Insulin
32 Crab
33 A cow
34 Vampire bats
35 Yoga
36 Rhesus
37 Canines
38 Cochlea
39 Hypnosis
40 Vectors

Quiz 4 (page 9)

41 46
42 Eight
43 9 m (30 ft)
44 12 pairs
45 31 pairs
46 R18
47 5-10 per cent
48 3.3 kg ($7\frac{1}{4}$ lb)
49 50 cm (20 in)
50 The person's age

Quiz 5 (page 9)

51 Laparoscopy
52 Biopsy
53 Defibrillator
54 Bleeding or bloodletting
55 Moxibustion
56 Homeopathy
57 Pain
58 Under the skin
59 Applied to the skin
60 Hypertension, or high blood pressure

Quiz 6 (page 9)

61 C
62 C
63 A
64 D
65 B
66 A
67 D
68 D
69 C
70 A

Quiz 7 (page 10)

71 Prescription
72 Limeys
73 Monosodium glutamate
74 Small or cramped spaces
75 Heartbeats
76 The feet
77 Vinegar and brown paper
78 Bath
79 Designer babies
80 Leeches

Quiz 8 (page 10)

81 Electrocardiogram, or electro-cardiograph
82 General practitioner
83 Intensive care unit
84 Intelligence quotient
85 Intrauterine device
86 In-vitro fertilisation
87 Measles, mumps and rubella
88 Multiple sclerosis
89 Repetitive strain injury
90 Sexually transmitted infection

Quiz 9 (page10)

91 The liver
92 50-60 per cent
93 Chromosomes
94 Pupil
95 Thigh and ear
96 Buttocks and ear
97 Alimentary canal
98 Bone marrow
99 Cervix
100 Sound, or hearing

Quiz 10 (page 10)

101 False
102 True
103 True
104 True
105 False
106 True
107 True
108 True
109 False
110 True

Quiz 11 (page 11)

111 'Deep Throat'
112 *King Lear*
113 Hypercusis
114 *Remembrance of Things Past*
115 *The Catcher in the Rye*
116 Tannochbrae
117 Holby General
118 *Fantastic Voyage*
119 Nanotechnology
120 Dendrimers

Quiz 12 (page 11)

121 Adrenaline
122 Lithium carbonate
123 Anastrazole
124 Aspirin
125 Aciclovir
126 Ranitidine ('Zantac')
127 Digoxin
128 Zidovudine (AZT)
129 Quinine
130 Chlorpromazine ('Largactil')

Quiz 13 (page 11)

131 D
132 B
133 B
134 D
135 B
136 C
137 A
138 D
139 C
140 D

Quiz 14 (page 12)

141 *One Flew Over the Cuckoo's Nest*
142 *The Millionairess*
143 'Goodness Gracious Me'
144 *The Singing Detective*
145 Arachnophobia
146 *The Scent of a Woman*
147 *Shine*
148 Blue Eyes
149 Obsessive-compulsive disorder
150 *Lorenzo's Oil*

Quiz 15 (page 12)

151 Nerve
152 Eczema
153 Thermometer
154 IUD
155 Pleurisy
156 Meningitis
157 Pulpitis
158 Osteoarthritis
159 Metastasis
160 Cyclosporin

Quiz 16 (page 12)

161 In the ear
162 Movement
163 Frequency of sound waves
164 Amplify it
165 1000 times
166 Muscles and tendons
167 Cold, heat, pain, pressure, vibration
168 Smell and taste
169 Monosodium glutamate (MSG)
170 Taste

Quiz 17 (page 12)

171 The Chinese
172 Gene or DNA amplification
173 Severe acute respiratory syndrome
174 Ayurveda, or Ayurvedic medicine
175 Smallpox
176 Sewers
177 Trepanning
178 Microsurgery
179 Genetic screening
180 Monoclonal antibodies

Quiz 18 (page 13)

181 F: Grandson/ Grandfather
182 D: Mother/ Daughter
183 E: Mother/Son
184 B: Son/Father
185 A: Brother/Brother
186 C: Brother/Brother
187 I: Brother/Sister
188 G: Sister/Sister
189 J: Sister/Sister
190 H: Father/Son

Quiz 19 (page 14)

191 False
192 False
193 True
194 True
195 False
196 False
197 True
198 True
199 True
200 False

Quiz 20 (page 14)

201 HIV/AIDS
202 Malaria
203 Meningitis
204 Pneumoconiosis, or silicosis
205 Schizophrenia
206 Laser
207 Hip replacement
208 Pass more urine
209 Insulin
210 Histamine

Quiz 21 (page 14)

211 Secondary sexual characteristics
212 Testosterone and oestrogen
213 Egg, or ovum
214 Adds fluid to semen
215 The penis
216 (a) Female, (b) Male
217 Testicular and prostate
218 35-40
219 FSH
220 Breast cancer and cervical cancer

Quiz 22 (page 14)

221 Iodine
222 Protein deficiency
223 Kos (or Cos)
224 Italy
225 Chelsea Physic Garden
226 The sauna
227 Zen
228 Meridians
229 Wormwood
230 (c); (d); (a); (b); (e)

Quiz 23 (page 15)

231 1980s
232 1810s
233 1960s
234 1890s
235 1940s
236 1950s
237 1800s
238 1910s
239 1980s
240 1940s

Quiz 24 (page 15)

241 Nucleus, mitochondria
242 The hyoid
243 Lymphocytes
244 The cerebellum
245 Lactic acid
246 The BCG vaccination against TB
247 *Chirurgia*
248 Stem cells
249 Karolinska Institute
250 Argentinian

Quiz 25 (page 15)

251 C
252 B
253 A
254 B
255 A
256 D
257 B
258 D
259 B
260 B

Quiz 26 (page 16)

261 The kidneys
262 The bladder
263 Joints
264 Teeth
265 The larynx, or voice box
266 5 litres (9 pints)
267 Iron
268 The stomach
269 Heart and lungs
270 Four

Quiz 27 (page 16)

271 Brimstone and treacle
272 Sulphur
273 Ear trumpet
274 Baldness
275 The nails
276 Colourings
277 Footprint, iris pattern, voice print
278 The bottom, or the feet
279 Pregnancy test
280 Stroke

Quiz 28 (page 16)

281 True
282 False
283 True
284 False
285 True
286 True
287 False
288 True
289 False
290 True

Quiz 29 (page 16)

291 Cod liver oil and orange juice
292 Yellow fever
293 Red Cross
294 Red and green
295 Yellow
296 Blood, black and yellow bile, phlegm
297 Red
298 Yellow spot
299 Reddish
300 Infrared or heat rays

Quiz 30 (page 17)

301 Rib
302 Thigh bone
303 Kneecap
304 Finger or toe bone
305 Spinal or back bone
306 Calf muscle
307 Windpipe
308 Voice box
309 Part of the small intestine
310 Gullet

Quiz 31 (page 17)

311 George Clooney, or Alex Kingston
312 *Dr Kildare*
313 Tony Hancock
314 *Cosby*
315 *Iris*
316 *Blackadder*
317 Captain Benjamin 'Hawkeye' Pierce
318 Mobile Army Surgical Hospital
319 4077th
320 Zaphod Beeblebrox

Quiz 32 (page 17)

321 Hodgkin's disease
322 The cornea
323 Growth hormone
324 Dementia
325 Narcolepsy
326 *Streptococcus*, or streptococcal
327 Carcinogens
328 Cerebral haemorrhage
329 Emphysema
330 Tissue rejection

Quiz 33 (page 17)

331 Wilhelm Roentgen
332 Robert Koch
333 Willem Einthoven
334 Thomas Hunt Morgan
335 Gerhard Domagk
336 Selman Waksman
337 Frederick Sanger
338 Godfrey Houndsfield
339 Kary Mullis
340 Stanley Prusiner

Quiz 34 (page 18)

341 Cystitis
342 Calcium
343 Chemotherapy
344 Cartilage
345 Corpuscles
346 Cholesterol
347 Constipation
348 Chloroform
349 Cradle cap
350 Cancers

Quiz 35 (page 18)

351 Mental illness
352 Smart bombs
353 Insulin
354 Ginseng
355 Breast cancer
356 A herbal
357 Willow, or meadowsweet
358 Opium poppy
359 Cinchona
360 Deadly nightshade

Quiz 36 (page 18)

361 Carbon, hydrogen and oxygen
362 Starch
363 Milk sugar and fruit sugar
364 Enzymes
365 Sugars
366 Fatty acids and glycerol
367 Detoxification
368 Urea
369 Its smell
370 The thyroid

Quiz 37 (page 18)

371 Frederick Sanger
372 Louis Pasteur
373 René Descartes
374 Christiaan Barnard
375 Jim Fixx
376 William Harvey
377 Sigmund Freud
378 Ivan Pavlov
379 Sir Alexander Fleming
380 Joseph Lister

Quiz 38 (page 19)

381 False
382 True
383 True
384 True
385 False
386 False
387 False
388 False
389 False
390 False

Quiz 39 (page 19)

391 Inside a cell
392 Cells
393 Macrophages
394 Hammer, anvil and stirrup
395 Bacteria
396 Pathogens
397 Minerals and vitamins
398 More
399 A virus
400 Hybridoma

Quiz 40 (page 19)

401 C
402 D
403 B
404 A
405 B
406 D
407 C
408 D
409 C
410 D

Quiz 41 (page 20)

411 Heart
412 Feet
413 Human liver
414 Lloyd Cole and the Commotions
415 '[They] only have eyes for you'
416 Joseph Conrad
417 'Dem [Dry] Bones'
418 The strings of my heart
419 Suzie
420 The Tintin books

Quiz 42 (page 20)

421 Pulmonary
422 Pheromones
423 Psychotherapy
424 Palpation
425 Prions
426 Porphyria
427 Potassium
428 Poliomyelitis
429 Porridge
430 Psalms

Quiz 43 (page 20)

431 1930s
432 Rabies
433 1977
434 14th century
435 1930
436 1984
437 About 1500
438 1960
439 16th century
440 Cataract

Quiz 44 (page 20)

441 Dentist
442 Systolic
443 Blood flow
444 Thermography
445 Endoscopy
446 MRI scanning
447 The heart
448 Deafness
449 It removes carbon dioxide
450 Lithotriptor

Quiz 45 (page 21)

451 Ovum cell
452 Skin cell
453 Red blood cell
454 Nerve cell
455 Fat cell
456 Liver cell
457 Cardiac muscle cell
458 Epithelial cells
459 White blood cell
460 Sperm cell

Quiz 46 (page 22)

461 Insomnia
462 Rubella, or German measles
463 Diabetes
464 Bacteria and viruses
465 Bone fractures
466 Lymph nodes
467 High blood pressure
468 Depression
469 38°C (100.4°F)
470 Poliomyelitis

Quiz 47 (page 22)

471 The cervix
472 Reflex
473 Water
474 Europe
475 Glaucoma
476 X-ray
477 Lung cancer
478 Transcendental meditation
479 Bach flower remedies
480 Drugs

Quiz 48 (page 22)

481 UK's smallest surviving baby
482 Siam
483 Stephen Hawking
484 Michelle Whitaker
485 The marathon
486 Julius Caesar
487 Arnold Schwarzenegger
488 Queen Victoria
489 Florence Nightingale
490 Nelson's blood

Quiz 49 (page 22)

491 True
492 True
493 False
494 True
495 True
496 False
497 True
498 False
499 True
500 True

Quiz 50 (page 23)

501 Soup noodles
502 Gastric juice and bile
503 Into the bloodstream
504 (a) Ureter, (b) Urethra
505 Ventricles
506 (a) Vitreous, (b) Aqueous
507 Into the bloodstream
508 The islets of Langerhans
509 Ink-blot patterns
510 Urine

Quiz 51 (page 23)

511 Melanocytes
512 Hair follicles
513 Carbon dioxide
514 Metabolism
515 Small intestine
516 Large intestine
517 Grey matter
518 Eustachian tube
519 Taste buds
520 Endocrine glands

Quiz 52 (page 23)

521 A
522 D
523 D
524 C
525 A
526 C
527 D
528 C
529 A
530 B

Quiz 53 (page 24)

531 Calcium
532 Fibre
533 Iron
534 The pancreas
535 Excretion
536 The kidneys
537 The liver
538 Fallopian tubes
539 Reflexes
540 Blood type or group

Quiz 54 (page 24)

541 NSAID
542 Bronchodilator
543 Cytotoxic drug
544 Laxative
545 Antipruritic
546 Antihistamine
547 Nitrates
548 ACE inhibitor
549 Hypnotic
550 Antifungal

Quiz 55 (page 24)

551 Attention deficit disorder
552 Body mass index
553 Bone-marrow transplant
554 Computerised axial tomography
555 Chronic obstructive pulmonary disease
556 Central nervous system
557 Electroconvulsive therapy
558 Electroencephalo-gram or -graph
559 Magnetic resonance imaging
560 Transient ischaemic attack

Quiz 56 (page 24)

561 Hygiene
562 Panacea
563 Psychology
564 Cognitive therapy
565 Lanugo
566 Vernix
567 The id
568 Superego
569 Front of the upper arm
570 The buttocks

Quiz 57 (page 25)

571 Foetal stethoscope
572 Stethoscope
573 Syringe
574 Scalpel
575 Ophthalmoscope
576 Ear thermometer
577 Laryngoscope
578 Clamping forceps
579 Dissecting forceps
580 Sphygmoman-ometer

Quiz 58 (page 26)

581 Garlic
582 Swab
583 Spread by contact or touch
584 Fungi
585 Tumour
586 A balloon
587 Crystal therapy
588 Sigmund Freud
589 Radiotherapy
590 Brain

Quiz 59 (page 26)

591 Boiling oil
592 The Hippocratic oath
593 Screening
594 On a prescription
595 The eyes
596 Cupping
597 Obstetrician
598 To heal cuts
599 A snake
600 Blood and bandages

Quiz 60 (page 26)

601 122
602 *Sleeping Beauty*
603 Cryonics
604 Scout
605 *What Ever Happened to Baby Jane?*
606 Henry Fonda and Katharine Hepburn
607 Jamie Bell
608 Zsa Zsa Gabor
609 Three times
610 Geriatrics

Quiz 61 (page 26)

611 T'ai Chi
612 The *haka*
613 The heartbeat
614 Vampires
615 Reflex action
616 Rowing, tennis, aerobics
617 Jazz dancing, squash, golf
618 Rapid eye movement
619 Causes contractions
620 Acupressure

Quiz 62 (page 27)

621 Acquired immune deficiency syndrome
622 Anabolism
623 Ammonia
624 Adrenal glands
625 Amniotic fluid
626 Antibodies
627 Arrhythmia
628 Actor
629 Anaesthetist, or anaesthesiologist
630 Anticoagulants, or anti-blood-clotting

Quiz 63 (page 27)

631 Orthopaedic surgeon
632 Epidemiologist
633 Pathologist
634 Urologist
635 Pharmacologist
636 Haematologist
637 Microbiologist
638 Biochemist
639 Endocrinologist
640 Otolaryngologist

Quiz 64 (page 27)

641 B
642 A
643 B
644 D
645 B
646 A
647 C
648 C
649 B
650 D

Quiz 65 (page 28)

651 Puberty
652 Phobia
653 Smoking
654 Kidney
655 The lungs, or bronchi
656 Stress
657 Hepatitis
658 Air pressure
659 Unconscious
660 Hair and nails

Quiz 66 (page 28)

661 Transfusion
662 Sigmund Freud
663 Heart transplant
664 Ultrasound scan
665 Osteopathy
666 Anaesthetic
667 Keyhole surgery
668 Hip replacement
669 Intravenous
670 Post mortem

Quiz 67 (page 28)

671 Immune system
672 Lymphatic system
673 Uvula
674 Pharynx
675 Tiny finger-like projections
676 The liver
677 The kidneys
678 Cortex
679 The pituitary
680 Cerebrospinal fluid

Quiz 68 (page 28)

681 True
682 False
683 True
684 False
685 True
686 True
687 False
688 False
689 True
690 True

Quiz 69 (page 29)

691 Nobel prize ceremony
692 Popeye
693 Lear
694 Nicolas Roeg
695 Jean Harlow
696 *Hamlet*
697 Rita Tushingham
698 Claude Rains
699 Paul Muni
700 Jane Horrocks

Quiz 70 (page 29)

701 Motor neuron disease
702 Tympanum
703 The labyrinth
704 Hormone system
705 Abnormal psychology
706 Pattern, form or shape
707 Naturopathy
708 Sleeping tablets
709 Thrombolytics
710 A pharmacopeia, or formulary

Quiz 71 (page 29)

711 B
712 A
713 C
714 C
715 B
716 A
717 C
718 B
719 D
720 D

Quiz 72 (page 30)

721 The stomach
722 Scrubbing up
723 Stitches
724 Stereoscopic
725 The heart
726 The skin
727 Nutrition
728 Whooping cough
729 Nerve or nervous
730 Fibre

Quiz 73 (page 30)

731 Gastroenteritis
732 Stomach ulcer
733 Kidney stone
734 Alzheimer's
735 Parkinson's
736 Mental illness
737 Glandular fever
738 Lyme disease
739 Scrofula
740 Secondaries

Quiz 74 (page 30)

741 Fever
742 'Infection'
743 The years
744 Scaffold
745 *Oliver Twist*
746 Fall in love
747 Martha and the Vandellas
748 A white Christmas
749 Radiohead
750 Madness

Quiz 75 (page 30)

751 Potassium
752 Antibodies
753 Antigen
754 Autonomic nervous system
755 Counter-irritation
756 Anaerobic
757 Reward and punishment
758 Yin and yang
759 An agonist
760 Antiviral

Quiz 76 (page 31)

761 Larynx
762 Retina
763 Aorta
764 Lining of the stomach
765 Lung
766 Fingernail
767 Bronchus
768 Surface of the tongue
769 Healthy heart
770 Brain folds

Quiz 77 (page 32)

771 True
772 False
773 False
774 False
775 True
776 True
777 True
778 True
779 False
780 False

Quiz 78 (page 32)

781 37°C
782 Haemoglobin
783 Mouth and nasal cavity or nose
784 The intercostal muscles
785 Blood
786 The prostate
787 Carbon dioxide
788 In the legs
789 The eyes
790 Placenta and umbilical cord

Quiz 79 (page 32)

791 Influenza epidemics
792 Sewers
793 Malaria
794 Michelangelo
795 Nicholas Culpeper
796 Beethoven
797 X-rays
798 Cholera
799 Intelligence tests
800 Hip replacement

Quiz 80 (page 32)

801 *Homo sapiens*
802 Histology
803 Cytoplasm
804 Heart attack
805 Roughage
806 Vegetables
807 Coeliac disease
808 Acupressure
809 Louis Pasteur
810 Pharmacology

Quiz 81 (page 33)

811 Martin Luther King
812 George Patton
813 Krebs cycle
814 Touch
815 Diabetes
816 Florence Griffith-Joyner
817 B.F. Skinner
818 Rudolf Steiner
819 Eyesight
820 Rosalind Franklin

Quiz 82 (page 33)

821 Achondroplasia
822 Apocrine glands
823 Autoimmune disease
824 Amino acids
825 Adrenal
826 Aphrodite
827 Atheromas
828 Autosuggestion
829 Anthroposophy
830 Aseptic surgery

Quiz 83 (page 33)

831 B
832 D
833 A
834 B
835 C
836 A
837 D
838 C
839 D
840 C

Quiz 84 (page 34)

841 Operating theatre
842 Sterilising instruments
843 Physiotherapy
844 Skin graft
845 Intensive care unit
846 Ultrasound
847 Scalpel
848 Open-heart surgery
849 Labour
850 Bedlam

Quiz 85 (page 34)

851 True
852 False
853 False
854 False
855 True
856 True
857 False
858 False
859 False
860 True

Quiz 86 (page 34)

861 Thrush, candidiasis, or candida
862 Botulism
863 Tetanus, or lockjaw
864 Diphtheria
865 Amoebic dysentery
866 Leprosy, or Hansen's disease
867 Meningitis
868 Typhoid
869 Cholera
870 Plague

Quiz 87 (page 34)

871 Tissues
872 Cowpox
873 Carotene
874 Marine algae, or seaweed
875 Contrast medium
876 Gene probe
877 Degenerative
878 Leukaemia and brain/nerve tumour
879 Plethora
880 Removal of a kidney or bladder stone

Quiz 88 (page 35)

881 Christiaan Barnard
882 Nettie Maria Stevens
883 Elizabeth Garrett Anderson
884 Sir Alexander Fleming
885 Christiane Nuesslein-Volhard
886 Dr Gertrude Elion
887 William Harvey
888 James Watson
889 Andreas Vesalius
890 Louis Pasteur

Quiz 89 (page 36)

891 Ingrowing toenails
892 Blood plasma
893 Vitamins
894 Spinal cord
895 Contractions
896 Emergency room
897 Aromatherapy
898 Vaccination
899 Melatonin
900 Antihistamine

Quiz 90 (page 36)

901 Vaccination
902 Reading glasses
903 Heart transplant
904 Aspirin
905 Cranial osteopathy
906 Epidural or spinal
907 Cobalt (-60)
908 Gamma rays, or electron beams
909 Poisons or kills cells
910 Organ comes from different species

Quiz 91 (page 36)

911 Gluten
912 The brain
913 Tinnitus
914 Profound deafness
915 Ectopic pregnancy
916 Warts
917 Height
918 Munchausen
919 Manic depression
920 Carcinoma

Quiz 92 (page 36)

921 Rhinitis
922 Bursitis
923 Osteoporosis
924 Tendinitis
925 Cirrhosis
926 Dialysis
927 Epiglottis
928 Bronchitis
929 Osteoarthritis
930 Meningitis

Quiz 93 (page 37)

931 True
932 True
933 True
934 False
935 True
936 False
937 True
938 True
939 False
940 True

Quiz 94 (page 37)

941 *Erin Brockovitch*
942 *Heartburn*
943 *Sex*
944 *Girl, Interrupted*
945 *The Big Sleep*
946 *Breathless*
947 *The Eyes of Laura Mars*
948 *Chariots of Fire*
949 *Insomnia*
950 *Maybe Baby*

Quiz 95 (page 37)

951 C
952 C
953 B
954 D
955 A
956 C
957 B
958 D
959 D
960 B

Quiz 96 (page 38)

961 False
962 True
963 True
964 False
965 True
966 True
967 True
968 True
969 False
970 True

Quiz 97 (page 38)

971 Uterus
972 Fontanelles
973 Christian Science
974 Biofeedback
975 Visualisation
976 The Great Stink
977 Transplant is living tissue; implant is not
978 Bionic
979 Respirator, or ventilator
980 Diphtheria

Quiz 98 (page 38)

981 Bruce Springsteen
982 The Boomtown Rats
983 Duran Duran
984 Elvis Presley
985 Orange Juice
986 Stevie Wonder
987 The Police
988 Michael Flanders and Donald Swann
989 Bing Crosby
990 Van Morrison

Quiz 99 (page 38)

991 Typhoid
992 Paralysis
993 Dengue fever
994 Tuberculosis
995 Oedema
996 Epilepsy
997 Influenza
998 Poliomyelitis
999 Arthritis
1000 Pneumonia

Question sheet

Quiz Number	Quiz Title

Questions	Answers
1	1
2	2
3	3
4	4
5	5
6	6
7	7
8	8
9	9
10	10

Answer sheet

Name

Quiz Number **Quiz Title**

Answers

1

2

3

4

5

6

7

8

9

10

Total score

F

G

H

I

Abbreviations:
T = top; M = middle; B = bottom; L = left; R = right

Front Cover: Getty Images/taxi
Back Cover: ©CORBIS/Rick Doyle

1 Martin Woodward. **2-3** Science Photo Library/BSIP, Laurent/B. Hop Ame. **4** Science Photo Library/CNRI. **13** Rex Features. **21** Top to bottom and left to right: Science Photo Library/John Giannicchi, 1; Science Photo Library/WG, 2, 4, 9, 10; Science Photo Library/Dr Tony Brain, 3; Science Photo Library/ISM, 5; Science Photo Library/Professors P.M. Motta, T. Fujita & M. Muto, 6; Science Photo Library/Don Fawcett, 7; Science Photo Library/J.W. Shuler, 8. **25** John Meek. **31** Top to bottom and left to right: Science Photo Library/Zephyr, 1; Science Photo Library/Paul Parker, 2; Science Photo Library/Sovereign, ISM, 3; Science Photo Library/Eye of Science, 4; Science Photo Library, 5, 9; Science Photo Library/Andrew Syred, 6; Science Photo Library/CNRI, 7; Science Photo Library/Prof. P. Motta/ Dept. of Anatomy/University of "La Sapienza", Rome; Science Photo Library/Geoff Tompkinson, 10. **35** Top to bottom and left to right: akg-images, 1; Carnegie Institute of Washington, 2; Bridgeman Art Library/Stapleton Collection, UK, 3; Science Photo Library/St. Mary's Hospital Medical School, 4; akg-images/Bruni Meya, 5; Science Photo Library/Will & Deni McIntyre, 6; Bridgeman Art Library/The Trustees of the Weston Park Foundation, UK, 7; Science Photo Library/A. Barrington Brown, 8; Bridgeman Art Library/Hermitage, St. Petersburg, Russia, 9; Bridgeman Art Library/ Private Collection, 10. **40-41** Getty Images/Phil Cole. **42** from, *Your Body Your Health: Genes and the Life Process* published by Reader's Digest ©2003/Rajeev Doshi, MM. **42-43** Science Photo Library/Quest, (background). **43** Science Photo Library/Quest, ML; Science Photo Library, TR; Science Photo Library/Ed Young, B. **44** Science Photo Library/CNRI, TL,TR; Science Photo Library/Gary Parker, BR. **44-45** Science Photo Library/Andrew Syred, M. **45** Bubbles/Angela Hampton, TR; SciencePhoto Library/Tek Image, (background). **46** Science Photo Library/Sheila Terry. **46-47** ©S.I.N./Corbis. **47** Science Photo Library/L'Oréal/Eurelios, TL,TM,TR; Stevie Williams, (faces), TL,TM TR; Laurence Bradbury, R. **48** Science Photo Library/Roger Harris, L; AntBits, BR. **49** Science Photo Library/James Stevenson, TL; Science Photo Library, R; Science Photo Library/Dr. Tony Brain, BL. **50** Martin Woodward, L: AntBits, MR. **51** Katz/Gamma, TM; Sporting Pictures, TR; Science Photo Library/Michael Abbey, BM. **52** Mirashade, M; Laurence Bradbury, BR. **53** Science Photo Library/CNRI, (background); AntBits, L; Science Photo Library/Schleichkorn/Custom Medical Stock Photo, MR. **54-55** Science Photo Library/Juergen Berger/Max-Planck Institute. **55** Science Photo Library/Andrew Syred, (background); Science Photo Library/Quest, M; Science Photo Library/Prof. P. Motta/Dept of Anatomy/University of "La Sapienza", Rome, MR; Science Photo Library/Dr Gopal Murti, BR. **56** Science Photo Library/Eye of Science, BL; AntBits, L. **56-57** Science Photo Library/Robert Becker/Custom Medical Stock Photo, (background). **57** Science Photo Library/Eye of Science, T; Science Photo Library/Petit Format/Institut Pasteur/Charles Dauguet, ML. **58** Science Photo Library/Innerspace Imaging. **59** Martin Woodward, T, ML; Science Photo Library/Dr Gary Settles, MR; Science Photo Library/Prof. P Motta/Dept. of Anatomy/University of "La Sapienza", Rome, B. **60** Martin Woodward. **61** Martin Woodward, T; Science Photo Library, BL; Bubbles/Jennie Woodcock, BR. **62** Science Photo Library/David Munns, MR; Science Photo Library/Gusto, BR. **62-63** Science Photo Library/Gusto, (background). **63** Science Photo Library/Sheila Terry, T; Science Photo Library/Martin Bond, BL. **64-65** John Meek. **66** Bubbles/Angela Hampton, TR. **66-67** Science Photo Library/Ed Young/Agstock, (background). **67** John Meek. **68** AntBits (main); from, *Your Body Your Health: The Stomach & Digestive System* published by Reader's Digest ©2001/Mirashade (inset). **69** Science Photo Library/ Eye of Science, T; Science Photo Library/Prof. P. Motta/Dept. of Anatomy/University of "La Sapienza", Rome, ML; Science Photo Library/CNRI, MR. **70** from, *Your Body Your Health: The Liver, Pancreas & Gallbladder* published by Reader's Digest ©2002/Mirashade. **71** from, *Your Body Your Health: The Kidneys and Urinary System* published by Reader's Digest ©2002/Mirashade, T; Martin Woodward, ML; Science Photo Library/CNRI, B; Science Photo Library, (background). **72** Mirashade. **73** Martin Woodward, T; Science Photo Library/Quest, B; Science Photo Library/David McCarthy, (background). **74** AntBits. **75** Science Photo Library/Sheila Terry, T; Science Photo Library/Alfred Pasieka, (background). **76** from, *Your Body Your Health: The Eyes & Mouth* published by Reader's Digest ©2003/Mirashade. **77** Martin Woodward, TL; Science Photo Library/Ralph Eagle, M; Science Photo Library/Will & Deni McIntyre, R; Science Photo Library/Sue Ford, B. **78** Martin Woodward, L; Science Photo Library, R. **79** Sporting Pictures/Tony Marshall, L; Science Photo Library/James King-Holmes, MR. **80** Science Photo Library/Eric Grave, T; AntBits, M; Science Photo Library/ Astrid & Hans-Frieder Michler, B. **81** Martin Woodward, TR; Science Photo Library/Prof. P. Motta/Dept. of Anatomy/University of "La Sapienza", Rome, TL; AntBits, B. **82** AntBits. **82-83** Science Photo Library/Alfred Pasieka, (background), **83** Katz/Piepenburg/laif, TR; Lucasfilm Ltd/Ronald Grant Archive, B. **84** Martin Woodward, R; Science Photo Library/Innerspace Imaging, TL; Science Photo Library/Prof. P.M. Motta, G. Macchiarelli, S.A. Nottola, BL. **85** AntBits, T; DigitalVision, M; Katz/Martin Black, BL. **86** Martin Woodward. **87** AntBits, T; Science Photo Library/Dept of Clinical Radiology, Salisbury District Hospital, (background). **88** from, *Your Body Your Health: The Reproductive System* published by Reader's Digest ©2003/Mirashade, TL; Science Photo Library/ Stevie Grand, BR. **89** Bubbles/Eddie Lawrance. **90** Bubbles/Pauline Cuttler, BL; Bubbles/Loisjoy Thurston, BR. **91** Martin Woodward, T; Science Photo Library/Philippe Rocher, BL. **92** ©Corbis/Rick Doyle, BL. **92-93** Science Photo Library/Mike Bluestone, (background), Rex Features/Brian Rasic, TR. **94** Bubbles/Grant Pritchard, B. **94-95** Rex Features, (background); Getty Images, R. **96** Science Photo Library/BSIP VEM, MR; Laurence Bradbury, B. **97** Science Photo Library/Mauro Fermariello. **98** Courtesy of the B.F. Skinner Foundation, B. **98-99** Laurence Bradbury after Rorshach, (background). **99** Science Photo Library/Bill Anderson, T. **100** Science Photo Library/John Greim. **100-101** Science Photo Library/John Greim, (background). **101** Rex Features. **102** Bridgeman Art Library/ Bibliothèque de la Faculté de Médecine, Paris, France/Archives Charmet, T; The Art Archive/Royal Library Stockholm/Dagli Orti, B. **102-3** Bridgeman Art Library/ Bibliothèque de la Faculté de Médecine, Paris, France/Archives Charmet (background). **103** The Art Archive/National Archaeological Museum, Athens/Dagli Orti, T; Bridgeman Art Library/Mauritshuis, The Hague, The Netherlands, M; Science Photo Library/James King-Holmes, B. **104** Science Photo Library/George Bernard, T; Science Photo Library/Patrick Donehue, BM. **104-5** Science Photo Library/ L. Steinmark/Custom Medical Stock Photo, (background). **105** Science Photo Library/Maximilian Stock, TR; Science Photo Library/Lunagrafix, (MR). **106** John Meek. **107** Bubbles/Frans Rombout, TL; John Meek, TR, M; Katz/Visum, BL. **108** Science Photo Library, BR; Science Photo Library/GJLP, BL. **109** Science Photo Library/Alfred Pasieka, TL; Science Photo Library/Sovereign, ISM, M; Science Photo Library/Tim Beddows, BL. **110** Science Photo Library/Professors P. Motta & T. Fujita/University of "La Sapienza", Rome. **110-111** Science Photo Library/Sinclair Stammers, (background). **111** Science Photo Library/Saturn Stills. **112** Ardea, London/Su Gooders, (background). **113** Science Photo Library/Edward Gray, MR; Science Photo Library/Eye of Science, (background). **114** Science Photo Library/Alfred Pasieka, M; Science Photo Library/NIBSC, B. **115** Science Photo Library/A.Gragera, Latin Stock, MR; Science Photo Library/CNRI, BL; Rex Features/Sipa, BR. **116** Katz/Gamma. **117** Science Photo Library/Quest, TR; Science Photo Library/BSIP VEM, M; Science Photo Library/Zephyr, BR. **118-119** Science Photo Library/Salisbury District Hospital. **119** Science Photo Library/CNRI, T; Science Photo Library/St Bartholomew's Hospital, M; Rex Features/M. Von Holden, BR. **120** Bridgeman Art Library/Private Collection/Archives Charmet. **121** Bridgeman Art Library/Linnean Society, London, TL; Bridgeman Art Library/Private Collection, BL; Bridgeman Art Library/ Alinari/ Osterreichische Nationalbibliothek, Vienna, Austria, BR. **122** Bridgeman Art Library/Roger-Viollet, Paris. **122-3** Science Photo Library/ BSIP/Chassenet, (background). **123** Bridgeman Art Library/Dinodia Picture Agency, Bombay, India, T; Katz /Bialobrzeski/laif, BL; Science Photo Library/Cristina Pedrazzini, BR. **124** Science Photo Library/Klaus Guldbrandsen, M; Science Photo Library, BL. **124-5** Science Photo Library/Chris Knapton, (background). **125** Science Photo Library/David Parker, MR; Science Photo Library/Françoise Sauze, B. **126** John Meek. **127** Bridgeman Art Library/ Archives Charmet/Private Collection, T; Science Photo Library/Jean-Loup Charmet, M; Katz/Meissner/laif, B. **128** Bridgeman Art Library/Musée Condé, Chantilly, France/ Lauros/Giraudon, MR; Bridgeman Art Library/Bibliothéque Nationale, Paris, France, BL. **129** Science Photo Library/Sheila Terry, TL; Bridgeman Art Library/Private Collection, M; The Art Archive/The Art Archive, BR. **130** Science Photo Library/James Stevenson, M. **130-1** Science Photo Library/Ed Young, (background). **131** Science Photo Library/BSIP, Laurent, TR; Science Photo Library/David M. Martin, M.D., M; Science Photo Library/BSIP, Edwige, BR. **132-3** Science Photo Library/Chris Bjornberg (background). **133** Science Photo Library/Catherine Pouedras/ MNHN/Eurelios, M; Science Photo Library/Ph. Plailly, Eurelios, (background). **134** Science Photo Library/BSIP, Leca, M; Science Photo Library/SIU., B. **135** Science Photo Library/Sheila Terry, T; Science Photo Library/BSIP, Laurent/B. Hop Ame, B. **136** Robert Harding Picture Library/Phototake, M; Martin Woodward, B. **136-7** Science Photo Library/Chris Priest & Mark Clarke, (background). **137** ©Corbis/Hulton-Deutsch Collection, TR; Science Photo Library/BSIP Boucharlat, ML; Science Photo Library/Claire Paxton & Jacqui Farrow, MB. **138-9** Science Photo Library/Colin Cuthbert, (background). **140** Science Photo Library/Dr Yorgos Nikas, M; Science Photo Library/W.A. Ritchie/Roslin Institute/Eurelios, B. **140-1** Science Photo Library/ Philippe Plailly, (background). **141** Science Photo Library/David Scharf, TM; Science Photo Library/W.A. Ritchie/Roslin Institute/Eurelios, MR. **142-3** Katz/Bialobrzeski/laif. **143** Colin Woodman, TL. **144** Science Photo Library. **145** Katz /Gamma.

Body and Health was published by The Reader's Digest Association Ltd, London. It was created and produced for Reader's Digest by Toucan Books Ltd, London.

Researched and written by
Michael Wright

For Toucan Books:
Editors
Jane Chapman, Helen Douglas-Cooper, Andrew Kerr-Jarrett
Picture researchers
Sandra Assersohn, Wendy Brown, Christine Vincent
Proofreader
Ron Pankhurst
Indexer
Dr Laurence Errington
Design
Bradbury and Williams

For Reader's Digest:
Project editor
Christine Noble
Project art editor
Louise Turpin
Pre-press accounts manager
Penny Grose

Reader's Digest, General Books:
Editorial director
Cortina Butler
Art director
Nick Clark

Colour origination
Colour Systems Ltd, London

Printed and bound
in Europe by Arvato, Iberia

ISBN 0 276 42937 0
BOOK CODE 625-005-01
CONCEPT CODE UK0095/G/S
ORACLE CODE 355500005H.00.24

First edition Copyright © 2004

The Reader's Digest Association Ltd,
11 Westferry Circus,
Canary Wharf,
London E14 4HE
www.readersdigest.co.uk

We are committed to both the quality of our products and the service we provide to our customers. We value your comments, so please feel free to contact us on 08705 113366 or via our web site at **www.readersdigest.co.uk**

If you have any comments or suggestions about the content of our books, you can email us at **gbeditorial@readersdigest.co.uk**